W0107458

Collegium Logicum

Annals of the Kurt-Gödel-Society

Volume 2

SpringerWienNewYork

Kurt-Gödel-Gesellschaft
Institut für Computersprachen
Technische Universität Wien, Vienna, Austria

Printing was supported by
Bundesministerium für Wissenschaft, Forschung und Kunst, Vienna

Printed on acid-free and chlorine-free bleached paper

With 3 Figures

ISSN 0947-577X
ISBN-13:978-3-211-82796-3 e-ISBN-13:978-3-7091-9461-4
DOI: 10.1007/ 978-3-7091-9461-4

Obituary

HAO WANG, 1921-1995

Prof. Hao Wang, Founding President of the Kurt Gödel Society, passed away on Saturday, May 13, 1995, in New York. The cause was lymphoma.
Professor Wang was one of the few confidants of Kurt Gödel, and a distinguished logician. He was one of the founding members of the Gödel Society and its first president, from 1987 to 1989. He is survived by his wife, a daughter, two sons, and two grandchildren.

Contents

Contents

Resolution Games and Non-Liftable Resolution Orderings

Hans de Nivelle

Department of Mathematics and Computer Science,
Delft University of Technology,
Julianalaan 132, 2628 BL, the Netherlands,
email: nivelle@cs.tudelft.nl

Abstract

We prove the completeness of the combination of ordered resolution and factoring for a large class of non-liftable orderings, without the need for any additional rules, as for example saturation. This is possible because of a new proof method which avoids making use of the standard ordered lifting theorem. This new proof method is based on a new technique, which we call the resolution game.

1 Introduction

Resolution is one of the most successful methods for automated theorem proving in first order classical logic. It was introduced in ([Robins65]). Shortly after its discovery it was realised that resolution could be improved by adding so called *refinements*. Refinements are restrictions of the resolution rule. An important group of refinements are the so called *ordering refinements*.
Refinements have to types of effect: They can improve the search efficiency in the case that a proof exists, and they can make the search process terminate in the case that a proof does not exist. The second may give rise to so called *resolution decision procedures*. (See [Joy76], [Zam72], [FLTZ93]). There is no hope for a general decision procedure, because provability in first order predicate logic is undecidable, but there exist some subclasses of formulae which are decidable.
In this paper we present a new class of orderings which can be used to define complete resolution refinements. They differ from the usual type of orders in the fact that they need not be invariant under substitutions. We will call this type of orders *non-liftable* orders. This makes it possible to define orderings that are much more total than the usual type of orderings. Using this type of orders we can expect more efficiency in proof search and more decision procedures.

We will first shortly describe how resolution and ordered resolution work, and then give the two completeness theorems for non-liftable orders.

Definition 1.1 We assume that there are a fixed, countable set V of variables, a fixed, countable set of function symbols, and a fixed, countable set of predicate symbols. We treat constants as functions that occur with arity 0.

The *terms* are the objects that can be obtained by finitely often applying the following rules: A variable is a term, and if f is a function symbol, and $t_1, \ldots, t_n$ are terms, then $f(t_1, \ldots, t_n)$ is a term. We do not assume that function symbols have a fixed arity.

If $t_1, \ldots, t_n$ are terms, and p is a predicate symbol, then $p(t_1, \ldots, t_n)$ and $\neg\, p(t_1, \ldots, t_n)$ are *literals.*

A *clause* is a finite set of literals. The *meaning* of a clause $c = \{A_1, \ldots, A_p\}$ equals $\forall v_1, \ldots, v_n(A_1 \vee \cdots \vee A_p)$, where $v_1, \ldots, v_p$ are the variables that occur in c.

The meaning of a clause is independent of the variables that are actually used. Because of this we will assume that variables in clauses can be replaced freely.

Definition 1.2 A *substitution* is a list which specifies how variables should be replaced. A substitution will have the form

$$\Theta = \{v_1 := t_1, \ldots, v_n := t_n\}.$$

This form prescribes that simultaneously all variables v_i have to be replaced by their corresponding t_i. In order to be meaningful it is necessary that $i \neq j \Rightarrow t_i \neq t_j$.

If A_1 and A_2 are literals, then A_1 is called an *instance* of A_2 if there is a substitution Θ, such that $A_1 = A_2\Theta$. The literals A_1 and A_2 are called *renamings* of each other if A_1 is an instance of A_2, and A_2 is an instance of A_1. Literal A_1 is called a *strict instance* of A_2 if A_1 is an instance of A_2, and A_1 and A_2 are not renamings.

If Θ_1 and Θ_2 are substitutions, then the *composition* of Θ_1 and Θ_2 is defined as follows:

$$\Theta_1 \cdot \Theta_2 = \{v := (v\Theta_1)\Theta_2 \mid (v\Theta_1)\Theta_2 \neq v\}.$$

If A_1 and A_2 are literals/atoms/terms, then a *unifier* is a substitution Θ, for which $A_1\Theta = A_2\Theta$. A *most general unifier* is a unifier Θ, such that for every unifier Σ there exists a substitution Ξ, such that $\Sigma = \Theta \cdot \Xi$.

Unifiers do always exist. This was first proven in ([Robins65]).

Theorem 1.3

1. If two literals/atoms/terms have a unifier then they have an mgu.
2. There exists an algorithm which computes the mgu of two literals, if it exists, and which reports failure otherwise.

We will first define an order, and then ordered resolution and factorisation.

Definition 1.4 First we define an *order*: An order is a relation $<$, with the following properties:

O1 Never $d < d$.

O2 If $d_1 < d_2$, and $d_2 < d_3$, then $d_1 < d_3$.

If S is a set then an element $s \in S$ is *maximal* in S if there is no $s' \in S$, with $s < s'$.

Next we define ordered resolution and factoring. Let $<$ be an order on literals:

Resolution Let $c_1 = \{A_1\} \cup R_1$ and $c_2 = \{\neg A_2\} \cup R_2$ be clauses (renamed, such that c_1 and c_2 have no overlapping variables), such that

1. A_1 is maximal in c_1, and $\neg A_2$ is maximal in c_2, and
2. A_1 and A_2 are unifiable,
3. Θ is the mgu of A_1 and A_2.

Then the clause $R_1\Theta \cup R_2\Theta$ is an *ordered resolvent* of R_1 and R_2.

Factorisation Let $c = \{A_1, \dots, A_p\} \cup R$ be a clause such that

1. one of the A_i is maximal in c,
2. the A_i are unifiable.

Then $c\Theta = \{A_1\Theta\} \cup R\Theta$ is an *ordered*-factor of c.

The resolution and factoring rule are applied as follows: If one wants to prove that a formula G is a logical consequence of formulae $F_1, \dots, F_p$ then the formula $F_1 \wedge \dots \wedge F_p \wedge \neg G$ is unsatisfiable. The formula $F_1 \wedge \dots \wedge F_p \wedge \neg G$ can be transformed into a finite set of clauses $C = \{c_1, \dots, c_n\}$ using standard techniques (see, for example, ([Lovelnd78]) or ([ChangLee73]). After this some fair search strategy has to be applied which will generate all possible resolvents and factors. Because the new derived clauses can be used in deriving new clauses, this process may last infinitely. If at any moment the empty clause is derived a proof has been found. Otherwise the search process has to continue.

Definition 1.5 Let $<$ be an order on literals. The order $<$ is *liftable* if

$$A < B \Rightarrow A\Theta \leq B\Theta.$$

The following theorem is the classical result that we want to improve:

Theorem 1.6 Let $<$ be a liftable order on literals. The combination of ordered resolution and factoring based on $<$ is complete.

Proofs can be found in ([Lovelnd78]), or ([ChangLee73]). According to ([Lovelnd78]), the idea of ordering clauses is from ([Reynolds66]).
The liftability condition imposes a rather strong restriction on the possible orders that can be used. For example a liftable order cannot compare $p(X)$ and $p(s(Y))$. If for example $p(X) < p(s(Y))$, then it would follow that $p(s(0)) < p(s(1))$, because of the substitution $\Theta_1 = \{X := s(0), Y = 1\}$. In the same way it is necessary that $p(s(1)) < p(s(0))$, because of the substitution $\Theta_2 = \{X := s(1), Y := 0\}$. This is impossible because $<$ is an order.
Our results allows to construct more total orderings, and to compare such pairs of literals:

Theorem 1.7 Let $<$ be an order on literals such that $A\Theta < A$ for all strict instances $A\Theta$ of A, and if $A < B$, then for all renamings $A\Theta_1$ of A, and $B\Theta_2$ of B, also $A\Theta_1 < B\Theta_2$. Then the combination of $<$-ordered resolution and factoring is complete.

Definition 1.8 Let $c = \{A_1, \ldots, A_p\}$ be a clause. c is called *decomposable* if for all literals A_i and A_j in c, either c_i and c_j have no shared variables, or c_i and c_j have exactly the same variables. The groups of literals which have the same variables are called *components*. A clause set C is called decomposable if all its members are decomposable.

The clause $\{p(X), q(Y), r(Y)\}$ is decomposable. The components are $\{p(X)\}$ and $\{q(Y), r(Y)\}$. The clause $\{p(X,Y), q(Y,Z)\}$ is not decomposable. The clause $\{p(X,Y), q(X,Y)\}$ is decomposable.

Theorem 1.9 Let $<$ be an order on literals such that $<$ satisfies the 2nd condition in Theorem 1.7. The combination of resolution and factoring is complete for decomposable clause sets.

The rest of this paper is dedicated to the proofs of these theorems.

2 Representation Indexed Clauses

The usual proof for the completeness of ordered resolution, for liftable orders in predicate logic, goes as follows: **(1)** If a clause set C is unsatisfiable, then there exists a finite set of ground instances which is unsatisfiable. **(2)** This set of ground instances has an ordered ground refutation, **(3)** and this ground refutation can be lifted to an ordered refutation of C. Step 3 is called the ordered *lifting* lemma/theorem. In the case of non-liftable orders, the ordered lifting lemma cannot be used anymore, because the non-ground clauses are probably ordered differently then the ground clauses. The following notion will help us to deal with this problem:

Definition 2.1 A *representation-indexed* clause is a clause of the form $\{a_1: A_1, \ldots, a_p: A_p\}$, such that there exists one substitution which makes for all i, $a_i = A_i\Theta$.

We will define resolution and factoring for representation-indexed clauses. We will write $r: R$ for a set of representation indexed-literals $\{c_1: C_1, \ldots, c_p: C_p\}$.

Resolution If $c_1 = \{a: A_1\} \cup r_1: R_1$ and $c_2 = \{\neg\, a: A_2\} \cup r_2: R_2$ are representation-indexed clauses, (with no overlapping variables) then the clause

$$r_1: R_1\Theta \cup r_2: R_2\Theta$$

is a resolvent of c_1 and c_2. Here Θ is the mgu of $\neg\, A_1$ and A_2.

Factorisation If $c = \{a: A_1, \ldots, a: A_p\} \cup r: R$ is a representation-indexed clause, then

$$\{a: A_1\Theta\} \cup r: R\Theta$$

is a factor of c. Here Θ is the mgu of $A_1, \ldots, A_p$.

Let $\overline{C}$ be a set of representation-indexed clauses. The set of *in-between* literals are the literals L, for which there is an indexed literal $a: A$ in $\overline{C}$, such that a is an instance of L, and L is an instance of A.

The in-between literals are the only interesting literals, because they are the only literals that can be produced by resolution and factoring. If $\overline{C}$ is a finite set of representation-indexed clauses, then the set of in-between literals of $\overline{C}$ that are not renamings of each other is finite.
The following proposition takes over a part of the lifting lemma.

Proposition 2.2 Definition 2.1 is meaningful.

This is because the necessary mgu will always exist, and the resulting clause will be a representation-indexed clause.
If a set of clauses C is unsatisfiable, then it is possible, using Herbrand's theorem, to construct a refutable set of representation-indexed clauses as follows:

Definition 2.3 If C is an unsatisfiable clause set, then we will write $\overline{C}$ for the following set of representation-indexed clauses:

1. Write $C = \{c_1, \ldots, c_n\}$. By Herbrand's theorem there exist substitutions $\Theta_{1,1}, \ldots, \Theta_{1,l_1}, \ldots, \Theta_{n,1}, \ldots, \Theta_{n,l_n}$ such that $\{c_1\Theta_{1,1}, \ldots, \Theta_{1,l_1}, \ldots, c_n\Theta_{n,1}, \ldots, \Theta_{n,l_n}\}$ is unsatisfiable, and ground.
2. For each $\Theta_{i,j}$, write c_i as $\{A_1, \ldots, A_p\}$. Then the representation-indexed clause

$$\{A_1\Theta_{i,j}{:}A_1, \ldots, A_p\Theta_{i,j}{:}A_p\} \in \overline{C}.$$

Proposition 2.4 Let $\overline{C}$ be a set of representation-indexed clauses obtained from a set of clauses C. Let $<$ be an order of the in-between literals of $\overline{C}$. Let $<'$ be an order on indexed $a{:}A$, for which a occurs in $\overline{C}$, and A is an in-between literal of $\overline{C}$. Assume that $A < B \Rightarrow a{:}A <' b{:}B$. If $\overline{C}$ has an ordered refutation using $<'$, then C has an ordered refutation using $<$.

Proof: The refutation of C can be obtained by replacing every clause $\{a_1{:}A_1, \ldots, a_p{:}A_p\}$ by $\{A_1, \ldots, A_p\}$ in the refutation of $\overline{C}$. □

Example 2.5 The following clauses are representation indexed: $\{p(0,1){:}p(X,Y), q(s(0),s(1)){:}q(s(X),s(Y))\}$, and $\{\neg\, p(0,1){:}p(A,1)\}$. A resolvent is $\{q(s(0),s(1)){:}q(s(X),s(1))\}$.

Now in order to prove that refutations of the type that is needed for Theorem 2.4 exist we will use a special tool called the resolution game.

3 The Resolution Game

The resolution game can be seen as a variant of lock resolution ([Boyer71]). Lock resolution is a resolution refinement in which natural numbers are attached to the literals. These natural numbers are called *indices.* The clauses are sorted, using the indices. Only literals which have the highest occurring index in the clause can be used for resolution. A clause that is obtained by resolution or factorisation inherits the indices from the parent clause(s). The resolution game is obtained from lock resolution by making three changes. The first two are not so fundamental, but the third is fundamental. First an arbitrary order is used on the indexed literals, instead of sorting the literals by index alone. Second the clauses will be multisets instead of sets. The third difference is that at every moment when a new clause is derived, a counterplayer may change the indices. In this way the deduction process becomes non-deterministic. We will prove that under certain limitations completeness is preserved.
Then, in the next section, we will apply the resolution game to the representation-indexed clauses of the previous section.

Definition 3.1 A (binary) resolution game $\mathcal{G}$ is a triple $\mathcal{G} = (P, \mathcal{A}, <)$, in which

- P is a finite set of literals,
- $\mathcal{A}$ is a set of attributes,
- $<$ is a well-order on $P \times \mathcal{A}$.

The elements of $P \times \mathcal{A}$ are called *indexed literals.* They are written as $a\!: A$, where $a \in P$, and $A \in \mathcal{A}$. A *clause* of $\mathcal{G}$ is a finite multiset of indexed literals of $\mathcal{G}$.

Definition 3.2 Let $\mathcal{G} = (P, \mathcal{A}, <)$ be a resolution game. We define:

Resolution Let $c_1 = [a\!: A_1] \cup R_1$ and $c_2 = [\neg\, a\!: A_2] \cup R_2$ be clauses of $\mathcal{G}$, such that $a\!: A_1$ is maximal in c_1, and $\neg\, a\!: A_2$ is maximal in c_2. Then $R_1 \cup R_2$ is an ordered resolvent of c_1 and c_2.

Factorisation Let $c = [a\!: A_1, a\!: A_2] \cup R$ be a clause of $\mathcal{G}$, such that $a\!: A_1$ is maximal in c. Then $c = [a\!: A_1] \cup R$ is an ordered factor of c.

Reduction: Let $c = [a_1\!: A_1, \ldots, a_p\!: A_p]$ be a clause of $\mathcal{G}$. A reduction of c is obtained by arbitrarily

1. deleting zero or more indexed literals from c.
2. and/or replacing zero or more indexed literals $a\!: A_i$ by an indexed literal $a\!: A_i'$, such that

$$a\!: A_i' < a\!: A_i.$$

Every clause is a reduction of itself. Every ordered factor of c is also a reduction of c. We will describe how the resolution game is played.

Definition 3.3 The resolution game has two players, the *defender* and the *opponent.* The opponent will try to derive the empty clause by using the ordered resolution rule, and the ordered factorisation rule. The defender will try to hinder the opponent by making reductions, and thus changing the order of the indexed literals.

The game starts with a set of initial clauses C, and proceeds as follows:

1. The defender first replaces all initial clauses by reductions. He has a completely free choice which deductions to take.
2. After this the opponent is allowed to derive ordered resolvents and ordered factors from the reduced clauses.
3. Before the opponent can use the new generation, the defender has the right to replace the new derived clauses by reductions. When this is finished the game continues at (2).

We will say that the opponent *wins* the resolution game if he derives the empty clause. The defender wins the resolution game if the empty clause is not derived. (In this case the resolution game may last infinitely long.)

Theorem 3.4 1. If the initial clause set is unsatisfiable, then there exists a winning strategy for the opponent.

2. If the initial clause set is satisfiable, then there exists a winning strategy for the defender.

It is not so difficult to see the second part of Theorem 3.4. When the defender never deletes a literal, the reductions are sound. Because ordered resolution and ordered factoring are sound rules, the empty clause will not be derived if the initial clause set is satisfiable.

4 Completeness of the Resolution Game

In this section we will prove the completeness of the resolution game. The proof that we give here is different from the proofs in ([Nivelle94a]) and ([Nivelle95]). We will use techniques similar to the techniques used in the Gentzen cut-elimination theorem. This technique exploits the relation between resolution and sequent calculi (see [Mints88]). We show that different applications of the resolution rule and the factorisation rule are interchangable. We start with an unordered resolution refutation tree of the initial clause set. During the game we rearrange the refutation in such a manner that the resolvents and factors will respect the ordering that is enforced by the defender. At the end of the game all clauses in the proof tree will be present in the game, and the empty clause will be derived. The same technique can also be used for proving the completeness of conventional types of ordered resolution.
A resolution refutation can be seen as a tree. The leaves of this tree are the initial clauses, and the root is the empty clause. We will define:

Definition 4.1 A *derivation tree* is a tree with labelled nodes. The nodes are labelled with multiset clauses. The labels must satisfy the following conditions:

- The leaves are labelled with initial clauses.
- The other nodes either have one successor or two successors. If a node has one successor then the label of this node is a factor of the label of its successor. If a node has two successors, then the label of this node is a resolvent of the labels of its successors.

The derivation tree is a *refutation tree* if the root equals the empty clause.

The refutation tree allows no structure sharing: When a clause is used more than one time, it occurs more than one time in the derivation tree.
We will now give the permutation rules. There are three types of permutations. The first type consists of permutations that permute two applications of the resolution rule. The second type consists of permutations that permute an application of the resolution rule with an application of the factorisation rule. The third type consists of the permutations that permute an application of the factorisation rule with an application of the factorisation rule.

Definition 4.2 The following permutations are possible: With A_1, A_2, we denote a complementary pair $a{:}\,A_1, \neg\, a{:}\,A_2$, or the converse $\neg\, a{:}\,A_1$, $a{:}\,A_2$. In the same way B_1, B_2 denotes a complementary pair of indexed literals:

RES

RES permutes two applications of the resolution rule:

$$\dfrac{\dfrac{[A_1, B_1] \cup R_1 \qquad [A_2] \cup R_2}{[B_1] \cup R_1 \cup R_2} \qquad [B_2] \cup R_3}{R_1 \cup R_2 \cup R_3}$$

permutes into

$$\dfrac{\dfrac{[A_1, B_1] \cup R_1 \qquad [B_2] \cup R_3}{[A_1] \cup R_1 \cup R_3} \qquad [A_2] \cup R_2}{R_1 \cup R_2 \cup R_3}$$

FACTRES1 The permutation FACTRES1 permutes an application of the resolution rule with an application of the factorisation rule, when the application of the factorisation rule is before the application of the resolution rule and the result of the factorisation rule is resolved upon by the resolution rule.

$$\dfrac{\dfrac{[A_1, A_1] \cup R_1}{[A_1] \cup R_1} \qquad [A_2] \cup R_2}{R_1 \cup R_2}$$

permutes into

$$\cfrac{\cfrac{\cfrac{\cfrac{[A_1, A_1] \cup R_1 \qquad [A_2] \cup R_2}{[A_1] \cup R_1 \cup R_2} \qquad [A_2] \cup R_2}{R_1 \cup R_2 \cup R_2}}{\cdots}}{R_1 \cup R_2}$$

Here one application of the factorisation rule is replaced by possibly many applications of the factorisation rule.

FACTRES2 The permutation FACTRES2 permutes an application of the resolution rule with an application of the factorisation rule, when the application of the factorisation rule is before the application of the resolution rule and the result of the factorisation rule is not resolved upon by the resolution rule.

$$\cfrac{\cfrac{[A, A] \cup [B_1] \cup R_1}{[A] \cup [B_1] \cup R_1} \qquad [B_2] \cup R_2}{[A] \cup R_1 \cup R_2}$$

permutes into

$$\cfrac{\cfrac{[A, A] \cup [B_1] \cup R_1 \qquad [B_2] \cup R_2}{[A, A] \cup R_1 \cup R_2}}{[A] \cup R_1 \cup R_2}$$

FACTRES3 Permutation FACTRES3 permutes an application of the resolution rule with an application of the factorisation rule, when the resolution rule is applied before the factorisation rule. It is the converse of FACTRES2.

FACT1 The permutation FACT1 permutes two applications of the factorisation rule, when the result of the first application is not factored upon by the second.

$$\cfrac{\cfrac{[A,A]\cup[B,B]\cup R}{[A]\cup[B,B]\cup R}}{[A]\cup[B]\cup R}$$

permutes into

$$\cfrac{\cfrac{[A,A]\cup[B,B]\cup R}{[A,A]\cup[B]\cup R}}{[A]\cup[B]\cup R}$$

FACT2 The permutation FACT2 permutes two applications of the factorisation rule when the result of the first application is again factored upon in the second application:

$$\cfrac{\cfrac{[A^1,A^2,A^3]\cup R}{[A^{12},A^3]\cup R}}{[A^{123}]\cup R}$$

permutes into

$$\cfrac{\cfrac{[A^1,A^2,A^3]\cup R}{[A^1,A^{23}]\cup R}}{[A^{123}]\cup R}$$

Theorem 4.3 All permutations of Definition 4.2 are correct, i.e., they transform correct resolution derivations into correct resolution derivations.

This can be easily checked. It is important to note that the rule FACTRES1 is the only rule in which the tree grows. (This implies that the art of designing complete resolution refinements lies in controlling the factorisation rule.) The permutation rules will be used as follows:

Lemma 4.4 Let C be an unsatisfiable clause set, that has a refutation tree. Assume that somewhere in the derivation tree there occurs a clause $c = [a\colon A, b\colon B] \cup R$, and that the indexed literal $a\colon A$ is used to derive the parent clause c, either by factorisation or by resolution.
Then it is possible to reorder the derivation tree, using the permutation rules of Definition 4.2 in such a manner that $b\colon B$ will be used in the derivation of the parent clause of c.

Proof: Somewhere higher in the tree (= closer to the root), the indexed literal $b\colon B$ will be used, because the root node is labelled with the empty clause. Then, using one of the rules of Definition 4.2, the use of $b\colon B$ can be moved one position closer to c. This process can be iterated until c is reached. □

Now imagine that we are opponent of an unsatisfiable indexed clause set C. We will proceed as follows: First we construct an unordered refutation tree for C, which uses indexed clauses. We will try to derive the clauses that occur in the tree, starting from the leaves and proceeding to the root. When a resolution step or factorisation step which we would like to make is blocked, we will reorder the tree, using Lemma 4.4. Sometimes, unfortunately, the number of nodes in the tree may increase due to an application of rule FACTRES1. In order to deal with this problem it is necessary to introduce the notion of *measure* of a resolution game situation.

Definition 4.5 Assume that the resolution game $\mathcal{G} = (P, \mathcal{A}, <)$ has arrived at a moment at which the opponent is on turn. We define the *measure* of the game situation. The measure either equals 0 or consists of an ordered pair $(a\colon A, n)$, where $a\colon A$ is an indexed literal and $n \in \mathcal{N}$. Let T be the part of the refutation tree, that is not yet derived in the resolution game.

- If there is no application of the factorisation rule in T, then the measure of the game situation equals 0.
- Otherwise $a\colon A$ is equal to the highest indexed literal upon which is factored in T, and n equals the number of times, that $a\colon A$ is factored upon.

We define that 0 is less than every measure of the form $(a\colon A, n)$. Measure $(a\colon A)$ is less then $(b\colon B, m)$ if $a\colon A < b\colon B$, or $a\colon A = b\colon B$, and $n < m$.

It is easily seen that the notion 'is less than' on measures is well-founded. When the refutation tree has to be rearranged we will do it in such a manner that the measure will decrease.
Assume that we are on turn. We proceed as follows:

1. Let c be a clause that occurs in the refutation tree. If there is a reduction c' of c that has been derived during the game, then we replace c by c' in the tree. This will be done for every clause for which this is possible. After this the tree will be adapted by adjusting the parents of the replaced clauses (because they inherit possibly different indices now, or indexed literals may have been deleted completely). During this process the measure will certainly not increase, because indexed literals are replaced by smaller indexed literals.

2. If it is possible to derive a new clause that is present in the refutation tree, but not in the game, then we will do this.

3. Otherwise it is necessary to reorder the tree. Let $a{:}A$ be the $<$-maximal indexed literal that occurs in the clauses of the refutation tree that are not yet derived. Then $a{:}A$ occurs in some clauses in the game. Using Lemma 4.4, we can reorder the tree in such a manner that $a{:}A$ will be used in the derivation of the parents of these clauses. During this reordering the measure of the game situtation will either remain the same or decrease because a factorisation on $a{:}A$ will be replaced by factorisation on a lower literal.

Because either the number of underived clauses in the tree decreases or the measure of the game situation decreases, we will in the end have derived the whole tree. As a consequence we will derive the empty clause.

5 Applications of the Resolution Game

In this section we will apply the resolution game in order to prove Theorem 1.7. If C is an unsatisfiable set of clauses we will construct an unsatisfiable representation-indexed clause set $\overline{C}$ from C. Then it is possible to construct a resolution game based $\overline{C}$ such that resolution and factoring as defined in Definition 2.1 will be a valid strategy of the defender.
In order to be able to apply the resolution game it is necessary to change it a little. In order to be able to do this we need the following notion:

Definition 5.1 Let $\preceq$ be a relation. The relation $\preceq$ is called a weak order if

O2' $d_1 \preceq d_2$ and $d_2 \preceq d_3$ implies $d_1 \preceq d_3$.

If $d_1 \preceq d_2$ and $d_2 \preceq d_1$, then we call d_1 and d_2 *equivalent*, notation $d_1 \equiv d_2$. A weak order is *well-founded* if the relation ($d_1 \preceq d_2$, and $d_1 \not\equiv d_2$) is well-founded.

We will adopt the notion of resolution game a little using this new notion of weak order.

Definition 5.2 We define the following variant of the resolution game:
A resolution game is defined as a triple $\mathcal{G} = (P, \mathcal{A}, \preceq)$, with

- P is a set of literals,
- $\mathcal{A}$ is a set of indices,
- $\preceq$ is a weak, well-founded order on $P \times \mathcal{A}$.

The game is played in essentially the same manner as in Section 3, but there are the following changes:

- A clause is defined as a set instead of a multiset.
- An indexed literal $a{:}A$ is considered maximal in a clause c if $a{:}A \in c$, and for no $b{:}B \in c$, it is the case that $a{:}A \preceq b{:}B$ and $a{:}A \not\equiv b{:}B$.
- Ordered resolvents and ordered factors are computed in the same way as in Definition 3.2, but using the new notion of maximality.
- A reduction of a clause c is obtained by replacing zero, one or more literals $a{:}A_1$ by a literal $a{:}A_2$ with $a{:}A_2 \preceq a{:}A_1$.

Theorem 5.3 If the game starts with an initial clause set C, and C is unsatisfiable, then there exists a winning strategy for the opponent.

This is easily seen, because equivalent indexed literals $a{:}A_1, a{:}A_2$ can be identified, and the result will be a resolution game in the sense of Definition 3.1.

Lemma 5.4 If $\overline{C}$ is a set of representation-indexed clauses, then every order $\prec$ which is invariant under renaming defines a weak, well-founded order $\preceq'$ on the set of in-between literals of $\overline{C}$ as follows:

$$A \preceq' B \text{ iff}$$

1. $A \prec B$, or
2. A and B are renamings of each other.

Proof: We show that $\prec'$ is reflexive. Always $A \preceq' A$, because A is a renaming of A. In order to show that $\preceq'$ is transitive assume that $A \prec' B$, and $B \prec' C$.

- If $A \prec B$, and $B \prec C$, then $A \prec C$, by transitivity of $\prec$. Then also $A \preceq' C$.
- If $A \prec B$, and B is a renaming of C, then $A \prec C$, because $\prec$ is invariant under renaming. As a consequence $A \preceq' C$.
- If A is a renaming of B, and $B \prec C$, then $A \prec C$, because $\prec$ is invariant under renaming. As a consequence $A \preceq' C$.

- If A is a renaming of B, and B is a renaming of C, then A is a renaming of C. Then also $A \preceq' C$.

It remains to show that $\preceq'$ is well-founded. This is the case, because there are only a finite number of non-equivalent in-between literals of $\overline{C}$. □

We will now prove the first completeness theorem for non-liftable orderings, Theorem 1.7. The proof will use the method suggested by Lemma 2.4. Let C be an unsatisfiable clause set. Construct a set $\overline{C}$ of representation-indexed ground instances of C, in the manner of Definition 2.3. Then construct the following resolution game of the type defined in Definition 5.3: Define $\mathcal{G} = (P, \mathcal{A}, \preceq)$, as follows:

- P is the set of ground literals that occur in $\overline{C}$.
- $\mathcal{A}$ equals the set of in-between literals of $\overline{C}$.
- $\preceq$ is defined from: $a{:}A \preceq b{:}B$ iff
  1. $A \prec B$, or
  2. A and B are renamings of each other.

The initial clause set equals $\overline{C}$. Now two things must be proven:

1. $\mathcal{G}$ is a resolution game, in the sense of Definition 5.3, and
2. resolution and factoring, as defined in Definition 2.1, is a valid strategy of the defender of $\mathcal{G}$.

In order to prove (1) it is sufficient to prove that $\preceq$ is a weak, well-founded order. This is the contents of Lemma 5.4.

(2) follows from the fact that in Definition 2.1 indices are always replaced by instances. These replacements are valid reductions because always $A\Theta \preceq A$.

We will end this section by giving a maximal order, which satisfies the conditions of Theorem 1.7. The order is maximal in the sense that it is not strictly included in another order which satisfies the conditions of Theorem 1.7. It defines an order between every pair of non-equivalent literals.

Definition 5.5 Assume that the variables are ordered as $V_0, V_1, \ldots, V_i, \ldots,$. We call a literal A is *standardised* if the following holds:

- If the variable V_{i+1} occurs in A, then the variable V_i occurs in A and, when A is written in the standard notation, every occurrence of V_{i+1} is preceded by an occurrence of V_i.

For example $p(f(V_1))$ is not standardised, but $p(f(V_0))$ is standardised. $q(V_1, V_0)$ is not standardised, but $q(V_0, V_1)$ is standardised.

Lemma 5.6 Every literal A has exactly one renaming, that is standardised. This literal is called the *standardisation* of A.

Definition 5.7 Let $\sqsubset$ be an order on predicate symbols (possibly negated), function symbols, and constant symbols. We first extend $\sqsubset$ to terms, and then to literals.

1. If f is a function symbol, or constant, and V is a variable, then $f \sqsubset V$.
2. For all variables V_i and V_j, we define
$$V_i \sqsubset V_j \text{ iff } i < j.$$
3. If f and g are n and m-ary function symbols (possibly with $n = 0$ or $m = 0$), and $f \sqsubset g$, then
$$f(t_1, \ldots, t_n) \sqsubset g(u_1, \ldots, u_m).$$
4. If f is an n-ary function symbol, and $f(t_1, \ldots, t_n) \neq f(u_1, \ldots, u_n)$, then let i the smallest integer, for which $t_i \neq u_i$. Then
$$f(t_1, \ldots, t_n) \sqsubset f(u_1, \ldots, u_n) \text{ iff } t_i \sqsubset u_i.$$

$\sqsubset$ is extended to literals by:

1. If f and g are (possibly negated) n and m-ary predicate symbols, and $f \sqsubset g$, then
$$f(t_1, \ldots, t_n) \sqsubset g(u_1, \ldots, u_m).$$
2. If f is an n-ary (possibly negated) predicate symbol, and $f(t_1, \ldots, t_n) \neq f(u_1, \ldots, u_n)$, then let i the smallest integer, for which $t_i \neq u_i$. Then
$$f(t_1, \ldots, t_n) \sqsubset f(u_1, \ldots, u_1) \text{ iff } t_i \sqsubset u_i.$$

Using this we define the following order $\sqsubset_s$ on literals: If A and B are literals then,
$$A \sqsubset_s B \text{ iff } A' \sqsubset B',$$
where A' is the standardisation of A, and B' is the standardisation of B.

Theorem 5.8 Let $\sqsubset$ be a total order on the (possibly negated) predicate symbols, function symbols, and constant symbols. Then the order $\sqsubset_s$, as defined in Definition 5.6, is an order that satisfies the conditions of Theorem 1.7.

As a consequence resolution and factoring with this order will be complete.

6 Decomposable Clauses

In this section we prove Theorem 1.9. Theorem 1.9 is especially important because many decidable classes are based on decomposable clause sets.
The fact that in Theorem 1.9 condition 2 of Theorem 1.7 can be dropped is caused by the fact that when a literal A increases its position in the ordering, during a substitution, all literals which have the same variables change. Because of this solidarity it is possible to have the other literals decrease, instead of A increase.
We will use the variant of the resolution game which is given in Definition 5.2. First note that sets of decomposable clauses are closed under resolution and factoring.
In order to be able to apply the resolution game it is necessary to adapt Lemma 2.4.

Lemma 6.1 Let C be an unsatisfiable set of clauses. Let $\overline{C}$ be obtained as in Definition 2.3. Let $\prec$ be an order on the in-between literals of $\overline{C}$. Let $\prec'$ be an order on the indexed literals $a{:}A$, for which a occurs in $\overline{C}$, and A is an in-between literal of $\overline{C}$.
For every pair $a{:}A, b{:}B$, for which there exist a substitution Θ and renaming Σ, such that $a = A\Theta$ and $b = B\Sigma\Theta$,

$$A < B \Rightarrow a{:}A <' b{:}B.$$

Then, if $\overline{C}$ has an ordered resolution refutation using $<'$, C has an ordered resolution refutation using $<$.

Proof: If one replaces all indexed clauses $\{a{:}A_1, \ldots, a_p{:}A_p\}$ in the refutation of $\overline{C}$ by $\{A_1, \ldots, A_p\}$, the result will be a refutation of C. □

Now the proof of Theorem 1.9 consists of two steps: **(1)** First the completeness will be proven in the case that C consists of one component clauses. **(2)** Second the full completeness will be proven. In order to prove step one the following resolution game will be used. Let C be an unsatifiable set of clauses that consist of one component. Let $\overline{C}$ be obtained as in Definition 2.3. We construct the following resolution game $\mathcal{G} = (P, \mathcal{A}, \preceq)$ of the type of Definition 5.2 as follows:

- P is the set of ground literals that occur in $\overline{C}$.
- $\mathcal{A}$ is the set of in-between literals of $\overline{C}$.
- $\preceq$ is defined as follows: Let $a_1{:}A_1$ and $a_2{:}A_2$ be indexed literals. Let Θ_1 be a $\subseteq$-minimal substitution such that $a_1 = A_1\Theta_1$, and let Θ_2 be a $\subseteq$-minimal substitution, such that $a_2 = A_2\Theta$. (The $\subseteq$-minimality is used to exclude redundant assignments.) Then $a_1{:}A_1 \preceq a_2{:}A_2$ if one of the following:

1. There is a substitution Σ_1, such that Θ_2 can be written as $\Theta_2 = \Sigma_1 \cdot \Theta_1$, but there is no substitution Σ_2, such that Θ_1 can be written as $\Theta_1 = \Sigma_2 \cdot \Theta_2$, or
2. there are substitutions Σ_1 and Σ_2, such that $\Theta_2 = \Sigma_1 \cdot \Theta_1$, and $\Theta_1 = \Sigma_2 \cdot \Theta_1$, and ($A_1 < A_2$, or A_1 and A_2 are renamings of each other.)

We must show that this is a correct resolution game. For this, it is sufficient to show that $\preceq$ is a weak, well-founded order. Reflexivity is immediate, transitivity follows from a case analysis. $\preceq$ is well-founded because the sets of non-equivalent in-between literals and non-equivalent in-between substitutions are finite. It is easily seen that resolution and factoring, as defined in Definition 6.1, is a correct strategy of the defender. It remains to prove the full completeness in the case that the clauses do not consist of one component. This can be proven by noticing that the different components in a clause are completely independent. Because of this they can be treated as literals. The whole process can then be seen as hyperresolution. □

With Theorem 1.9, an open question in ([FLTZ93]) can be solved. Before we give it we need some definitions.

Definition 6.2 A term is *functional* if it is of the form $f(t_1, \ldots, t_n)$. A literal A is called *weakly covering* if every (sub)term of A, that is non-ground and functional, contains all variables that occur in A.

Definition 6.3 Let C be a set of clauses. C is in E^+ iff

1. All literals occurring in clauses of C are weakly covering, and
2. all clauses in C are decomposable.

Definition 6.4 The $<_v$ order is the following order on literals: $A_1 <_v A_2$ iff the maximal depth of occurrence of a variable in A_1 is strictly less then the maximal depth of occurrence of a variable in A_2.

It is proven in ([FLTZ93]) that, when resolution with factoring is applied on a set of clauses in the E^+-class, then only a finite set of clauses will be derived. However the completeness of resolution with the $<_v$-order was open. The completeness of resolution with the $<_v$-order on the E^+-class follows from Theorem 1.9. So, now it is proven that resolution and factoring with the $<_v$-order is a decision procedure for clause sets that are in E^+.

References

[BG90] L. Bachmair, H. Ganzinger, On restrictions of ordered paramodulation with simplification, CADE 10, pp 427-441, Kaiserslautern, Germany, Springer Verlag, 1990.

[Boyer71] R.S. Boyer, Locking: A Restriction of Resolution, Ph. D. Thesis, University of Texas at Austin, Texas 1971.

[ChangLee73] C-L. Chang, R. C-T. Lee, Symbolic logic and mechanical theorem proving, Academic Press, New York 1973.

[Egly] On Definitional Transformations to Normal Form for Intuitionistic Logic, unpublished.

[FLTZ93] C. Fermüller, A. Leitsch, T. Tammet, N. Zamov, Resolution Methods for the Decision Problem, Springer Verlag, 1993.

[Galllier86] Jean H. Gallier, Logic for Computer Science, (Foundations of Automatic Theorem Proving), Harper & Row, Publishers, New York, 1986.

[Girard87] Linear Logic, in Theoretical Computer Science 50, pages 1-102, 1987.

[HR91] J. Hsiang and M. Rusinowitch, Proving Refutational Completeness of Theorem-Proving Strategies: The Transfinite Semantic Tree Method, Journal of the ACM, Vol. 38, no. 3, July 1991, pp. 559-587.

[Joy76] W.H. Joyner, Resolution Strategies as Decision Procedures, J. ACM 23, 1 (July 1976), pp. 398-417.

[KH69] R. Kowalski, P.J. Hayes, Semantic trees in automated theorem proving, Machine Intelligence 4, B. Meltzer and D. Michie, Edingburgh University Press, Edingburgh, 1969.

[Leitsch88] A. Leitsch, On Some Formal Problems in Resolution Theorem Proving, Yearbook of the Kurt Gödel Society, pp. 35-52, 1988.

[Loveland78] D. W. Loveland, Automated Theorem Proving, a Logical Basis, North Holland Publishing Company, Amsterdam, New York, Oxford, 1978.

[Mints88] G. Mints, Gentzen-Type Systems and Resolution Rules, Part 1, Propositional Logic, in COLOG-88, International Conference on Computational Logic, Talinn (at that time) USSR, 1988.

[Nivelle94a] H. de Nivelle, Resolution Games and Non-Liftable Resolution Orderings, Internal Report 94-36, Department of Mathematics and Computer Science, Delft University of Technology, 1994.

[Nivelle94b] H. de Nivelle, Application of Resolution Games to Resolution Decision Procedures, Internal Report 94-50, Department of Mathematics and Computer Science, Delft University of Technology, 1994.

[Nivelle95] Resolution Games and Non-Liftable Resolution Orderings, in CSL 94, pp 279-293, Springer Verlag, Kazimierz, Poland, 1994,

[Reynolds66] J. Reynolds, Unpublished Seminar Notes, Stanford University, Palo Alto, California, 1966.

[Robins65] J.A. Robinson, A Machine Oriented Logic Based on the Resolution Principle, Journal of the ACM, Vol. 12, pp 23-41, 1965.

[Stat79] R. Statman, Lower Bounds on Herbrand's Theorem, in Proceedings of the American Mathematical Society, Vol. 75, Number 1, 1979.

[Tamm93] T. Tammet, Proof Search Strategies in Linear Logic, Report 70, Programming Methodology Group, Department of Computer Sciences, Chalmers University of Technology and University of Göteburg, 1993.

[Tamm94] T. Tammet, Separate Orderings for Ground and Non-Ground Literals Preserve Completeness of Resolution, unpublished, 1994.

[Zam72] N.K. Zamov: On a Bound for the Complexity of Terms in the Resolution Method, Trudy Mat. Inst. Steklóv 128, pp. 5-13, 1972.

A Tableau Calculus for Partial Functions*

Manfred Kerber Michael Kohlhase

Fachbereich Informatik, Universität des Saarlandes
66041 Saarbrücken, Germany
{kerber|kohlhase}@cs.uni-sb.de
URL: http://jswww.cs.uni-sb.de/

Abstract. Even though it is not very often admitted, partial functions do play a significant role in many practical applications of deduction systems. Kleene has already given a semantic account of partial functions using a three-valued logic decades ago, but there has not been a satisfactory mechanization. Recent years have seen a thorough investigation of the framework of many-valued truth-functional logics. However, strong Kleene logic, where quantification is restricted and therefore not truth-functional, does not fit the framework directly. We solve this problem by applying recent methods from sorted logics. This paper presents a tableau calculus that combines the proper treatment of partial functions with the efficiency of sorted calculi.

Keywords: Partial functions, many-valued logic, sorted logic, tableau.

1 Introduction

Many practical applications of deduction systems in mathematics and computer science rely on the correct and efficient treatment of partial functions. For this purpose different approaches—reaching from workarounds for concrete situations to a proper general treatment—have been developed. In the following we will introduce the main approaches and exemplify their advantages and disadvantages by some trivial examples from arithmetic. For a more detailed discussion of the different approaches compare [Far90].

There are essentially four approaches of treating partiality. First, these expressions can syntactically be excluded. Second, it is possible to disregard or bypass partiality. Third, partiality is taken serious and this is reflected in the semantics and the calculus. Fourth, there is some mixture between options two and three.

In the first approach terms like $\frac{x}{0}$ are treated as syntactically ill-formed, for instance, by using a sorted logic, in which the domain of the $\frac{x}{y}$ function is defined to be $\mathbb{R} \times \mathbb{R}^*$ (where $\mathbb{R}^*$ denotes the real numbers without 0). Thereby the whole problem of partiality has been bypassed. In the cases, where such a procedure is possible, this approach is quite adequate and reflects the usual way

* This work was supported by the Deutsche Forschungsgemeinschaft (SFB 314, D2)

of handling undefined expressions in mathematics: to assure that all expressions are defined before beginning to reason about them. It is, however, not always possible to exclude such expressions from the consideration a priori. For instance, if you consider terms like $\frac{1}{f(x)}$, it would be necessary to exclude this expression for those x where $f(x) = 0$; depending on the definition of f, this might be not computable at all. In consequence, this approach remedies the problem of partiality in certain cases only and does not provide a full solution.

In the second approach a value is assigned to $\frac{1}{0}$, either a fixed value (e.g. 0) or an undetermined one. In both cases it is necessary to tolerate undesired theorems, in the first case, for instance, $\frac{1}{0} = 0$, or in the second case from $0 \cdot x = 0$ the instance $0 \cdot \frac{1}{0} = 0$. This approach is not satisfying, if such theorems are unwanted, which is normally the case in mathematics.

In the third approach, terms like $\frac{1}{0}$ are not defined and semantically either uninterpreted or interpreted by some error element. In the same manner, atomic formulae, containing such an undefined term, like $\frac{1}{0} = 0$ are not interpreted by a truth value (true or false) at all or are interpreted by a third truth value (undefined). As in the first approach, partiality is taken serious, but it is no longer necessary to single out the undefined expressions a priori. The main drawback of this approach is that classical two-valued logic is not adequate for its mechanization. A possible formalization can be done by a three-valued logic, however. Kleene makes this approach formal, by introducing an individual $\bot$ denoting meaningless individuals and a third truth value u, standing for the "undefined" truth value. However, in contrast to the general framework for many-valued truth-functional logics, Kleene's quantifiers only range over defined values, that is, not over $\bot$, making a direct utilization of the methods developed by Carnielli [Car87, Car91], Hähnle [Häh92], Baaz and Fermüller [BF92] impossible. Kleene's approach has been used by Tichy [Tic82], Lucio-Carrasco and Gavilanes-Franco [LCGF89] to give logical systems for partial functions. Both approaches offer unsorted operationalizations of the systems in sequent calculi.

The fourth approach is less radical insofar as terms are treated as in the third approach, but the problems that accompany treating a third truth value are avoided (cf. [Bee85, Far90, Sch68, Wei89]): All atomic expressions containing a meaningless term are considered as false. This has the advantage that partial functions can be handled within the classical two-valued framework. However, the serious drawback is that the results of these logic systems can be unintuitive to the working mathematician. For instance in elementary arithmetic the following sentence

$$\forall x, y, z.\, z = \frac{x}{y} \Rightarrow x = y * z$$

is a theorem of such systems since the scope is true for the case $y \neq 0$ and for the case $y = 0$, the formula $z = \frac{x}{0}$ obtains the truth value f which in turn makes the implication true, too. However, it is mathematical consensus that the equation should only hold provided that y is not 0. It will turn out (cf. example 2.11) that the formula is not a theorem in our formalization, since the case $y = 0$ is a counterexample.

This paper formalizes Kleene's ideas for partial functions (the third approach) in a sorted three-valued logic, called $\mathcal{SKL}$, that uses Kleene's strong interpretation of connectives and quantifiers and adapts techniques from Weidenbach's sorted logic [Wei89] to handle definedness information. We furthermore present a tableau calculus $\mathcal{TPF}$ for partial functions that carries over the methods developed in the context of resolution theorem proving for partial functions [KK94] to the tableau framework. Standard first-order tableaux were introduced by Beth [Bet55] and Hintikka [Hin55] and later unified by Smullyan [Smu68]. The free variable tableau method has its origin in the work of Prawitz [Pra60] and has further been elaborated by Reeves [Ree87] and Fitting [Fit90]. Both calculi reported here are strongly influenced by Weidenbach's tableau calculus with sorts [Wei94], which introduces reasoning with dynamic sorts to tableau calculi.

We would like to thank Christian Fermüller, Reiner Hähnle, and Christoph Weidenbach for comments and clarifying discussions.

2 Strong Sorted Kleene Logic ($\mathcal{SKL}$)

In [Kle52] Kleene presents a logic, which he calls *strong three-valued logic* for reasoning about partial recursive predicates on the set of natural numbers. He argues that the intuitive meaning of the third truth value should be "undefined" or "unknown" and introduces the truth tables shown in definition 2.6. Similarly Kleene enlarges the universe of discourse by an element $\bot$ denoting the undefined number. In his exposition the quantifiers only range over natural numbers, in particular he does not quantify over the undefined individual (number).

The approach of this paper is to make Kleene's meta-level discussion of defined and undefined individuals explicit by structuring the universe of discourse with the sort $\mathfrak{D}$ for all defined individuals. Furthermore all functions and predicates are strict, that is, if one of the arguments of a compound term or an atom evaluates to $\bot$, then the term evaluates to $\bot$ or the truth value of the atom is u. Just as in Kleene's system, our quantifiers only range over individuals in $\mathfrak{D}$, that is, individuals that are not undefined. This is in contrast to the well-understood framework for truth-functional many-valued logics, where the concept of definedness and defined quantification cannot be easily introduced, since quantification is truth-functional and depends on the truth values for all (even the undefined) instantiations of the scope. Kleene's concept of bounded quantification is essential for our program of representing partial functions, since in a truth-functional approach no proper universally quantified expression can evaluate to the truth value t (dually for the existential quantifier), since all functions and predicates are assumed strict.

In the following we present the logic system $\mathcal{SKL}$, which is a sorted version of what we believe to be a faithful formalization of Kleene's ideas from [Kle52]. We treat the sorted version here, since we need the machinery for dynamic sorts in the calculus to be able to treat the sort $\mathfrak{D}$ (sort techniques as that from [Wei89, Wei91] give us the bounded quantification). We will call formulations of $\mathcal{SKL}$ where $\mathfrak{D}$ is the only sort in the signature *strong unsorted Kleene logic*, since the sort $\mathfrak{D}$ is indispensable. The further use of sorts gives the well-known advantages of sorted logics for the conciseness of the representation and the reduction of search spaces.

2.1 Syntax and Semantics

Definition 2.1 (Signature) A *signature* $\Sigma := (\mathcal{S}, \mathcal{V}, \mathcal{F}, \mathcal{P})$ consists of the following disjoint sets

- $\mathcal{S}$ is a finite set of *sorts* including the sort $\mathfrak{D}$. We define $\mathcal{S}^* := \mathcal{S} \setminus \{\mathfrak{D}\}$.
- $\mathcal{V}$ is a set of *variable symbols.* Each variable x is associated with a unique sort S, which we write in the index, i.e. x_S. We assume that for each sort $S \in \mathcal{S}$ there is a countably infinite supply of variables of sort S in $\mathcal{V}$.
- $\mathcal{F}$ is a set of *function symbols.*
- $\mathcal{P}$ is the set of *predicate symbols.*

The sets $\mathcal{F}$ and $\mathcal{P}$ are subdivided into the sets $\mathcal{F}^k$ of *function symbols of arity k* and $\mathcal{P}^k$ of *predicate symbols of arity k*. Note that individual constants are just nullary functions. We call a signature *unsorted* if $\mathcal{S}^*$ is empty, that is, if $\mathfrak{D}$ is the only sort.

Definition 2.2 (Terms and Formulae) We define the set of *terms* to be the set of variables together with *compound terms* $f(t^1, \ldots, t^k)$ for terms $t^1, \ldots, t^k$ and $f \in \mathcal{F}^k$.

If $P \in \mathcal{P}^k$, then $P(t^1, \ldots, t^k)$ is a *proper atom.* If t is a term and S a sort then $t \lessdot S$ is a *sort atom.* The set of *formulae* contains all atoms and with formulae A and B the formulae $A \wedge B$, $A \vee B$, $A \Rightarrow B$, $\neg A$, $!A$, $\forall x_S.\ A$, and $\exists x_S.\ A$.[1] Here the intended meaning of $!A$ is that A is defined.

We will now define the three-valued semantics for $\mathcal{SKL}$ by postulating an "undefined individual" $\perp$ in the universe of discourse. Note that this is similar to the classical flat CPO construction [Sco70], but Kleene's interpretation of truth values does not make u minimal. Since we are not interested in least fixpoints, monotonicity does not play a role in this paper.

Definition 2.3 (Strict Σ-Algebra) Let Σ be a signature, then a pair $(\mathcal{A}, \mathcal{I})$ is called a *strict Σ-algebra*, iff

1. the *carrier set* $\mathcal{A}$ is an arbitrary set that contains $\perp$,
2. the *interpretation function* $\mathcal{I}$ obeys the following restrictions:
   (a) For all function symbols f, the function $\mathcal{I}(f): \mathcal{A}^k \longrightarrow \mathcal{A}$ is strict for $\perp$, that is, $\mathcal{I}(f)(a_1, \ldots, a_k) = \perp$, if $a_i = \perp$ for (at least) one i.
   (b) If P is a predicate symbol, then the relation $\mathcal{I}(P) \subseteq \mathcal{A}^k$ is strict for $\perp$, that is, $\mathcal{I}(P)(a_1, \ldots, a_k) = \mathsf{u}$, if $a_i = \perp$ for (at least) one i.

[1] We do not consider degenerate quantifications of the form $\forall x_S.\ A$, where x does not occur free in A, they would require a special treatment in the calculus. For a treatment in the resolution framework see [KK93, Remark 3.3 and 3.4].

(c) If $S \neq \mathfrak{D}$ is a sort, then $\mathcal{I}(S)$ is a total and strict unary relation, that is, $\mathcal{I}(S)(a) \in \{\mathsf{f},\mathsf{t}\}$, if $a \neq \perp$ and $\mathcal{I}(S)(\perp) = \mathsf{u}$.

(d) $\mathcal{I}(\mathfrak{D})(\perp) = \mathsf{f}$ and $\mathcal{I}(\mathfrak{D})(a) = \mathsf{t}$, if $a \neq \perp$. Note that in contrast to all other sorts and predicates, the denotation of $\mathfrak{D}$ is not a strict relation.

We define the *carrier* $\mathcal{A}_S$ *of sort* S as $\mathcal{A}_S := \{a \in \mathcal{A} \mid \mathcal{I}(S)(a) = \mathsf{t}\}$. Note that in contrast to other sorted logics, it is not assumed that the $\mathcal{A}_S$ are non-empty, in fact we do not even assume the existence of defined elements in the carrier. Furthermore $\perp \notin \mathcal{A}_S$ for any $S \in \mathcal{S}$.

By systematically deleting $\perp$ and u from the carrier and the truth values we can canonically transform strict Σ-algebras into algebras of partial functions. These are an algebraic account of the standard interpretation in mathematics, where partiality of functions is directly modeled by right-unique relations. Obviously these notions of algebras have a one-to-one correspondence, so both approaches are equivalent.

Definition 2.4 (Σ-Assignment) Let $(\mathcal{A},\mathcal{I})$ be a strict Σ-algebra, then we call a total mapping $\varphi\colon \mathcal{V} \longrightarrow \mathcal{A}$ a Σ-*assignment*, iff $\varphi(x_S) \in \mathcal{A}_S$, provided $\mathcal{A}_S$ is non-empty and $\varphi(x_S) = \perp$ if $\mathcal{A}_S = \emptyset$. We denote the Σ-assignment that coincides with φ away from x and maps x to a with $\varphi,[a/x]$.

Definition 2.5 Let φ be a Σ-assignment into a strict Σ-algebra $(\mathcal{A},\mathcal{I})$ then we define the *value function* $\mathcal{I}_\varphi$ *from formulae to* $\mathcal{A}$ inductively to be

1. $\mathcal{I}_\varphi(f) := \mathcal{I}(f)$, if f is a function or a predicate.
2. $\mathcal{I}_\varphi(x) := \varphi(x)$, if x is a variable.
3. $\mathcal{I}_\varphi(f(t^1,\ldots,t^k)) := \mathcal{I}(f)(\mathcal{I}_\varphi(t^1),\ldots,\mathcal{I}_\varphi(t^k))$, if f is a function or predicate.
4. $\mathcal{I}_\varphi(t \lessdot S) := \mathcal{I}(S)(\mathcal{I}_\varphi(t))$.

Since this definition applies to $\mathcal{P}$ and $\mathcal{F}$ alike, we have given the semantics of all atomic formulae. The semantic status of sorts is that of total unary predicates; in particular we have $\mathcal{I}_\varphi(t \lessdot S) = \mathsf{u}$, iff $\mathcal{I}_\varphi(t) = \perp$ for $S \neq \mathfrak{D}$.

Definition 2.6 The value of a formula dominated by a connective is obtained from the value(s) of the subformula(e) in a truth-functional way. Therefore it suffices to define the truth tables for the connectives:

∧	f	u	t
f	f	f	f
u	f	u	u
t	f	u	t

∨	f	u	t
f	f	u	t
u	u	u	t
t	t	t	t

⇒	f	u	t
f	t	t	t
u	u	u	t
t	f	u	t

¬	
f	t
u	u
t	f

!	
f	t
u	f
t	t

The semantics of the quantifiers is defined with the help of function $\widetilde{\forall}$ and $\widetilde{\exists}$ from the non-empty subsets of the truth values in the truth values. We define

$$\mathcal{I}_\varphi(\mathsf{Q}x_S.\, A) := \widetilde{\mathsf{Q}}(\{\mathcal{I}_{\varphi,[a/x]}(A) \mid a \in \mathcal{A}_S\})$$

where $Q \in \{\forall, \exists\}$ and furthermore

$$\tilde{\forall}(T) := \begin{cases} \mathsf{t} & \text{for } T = \{\mathsf{t}\} \text{ or } T = \emptyset \\ \mathsf{u} & \text{for } T = \{\mathsf{t},\mathsf{u}\} \text{ or } \{\mathsf{u}\} \\ \mathsf{f} & \text{for } \mathsf{f} \in T \end{cases} \qquad \tilde{\exists}(T) := \begin{cases} \mathsf{t} & \text{for } \mathsf{t} \in T \\ \mathsf{u} & \text{for } T = \{\mathsf{f},\mathsf{u}\} \text{ or } \{\mathsf{u}\} \\ \mathsf{f} & \text{for } T = \{\mathsf{f}\} \text{ or } T = \emptyset \end{cases}$$

Note that with this definition quantification is separated into a truth-functional part $\tilde{\forall}$ and an instantiation part that only considers members of $\mathcal{A}_S$. Since $\bot$ is not a member of any $\mathcal{A}_S$, quantification never considers it and therefore cannot be truth-functional even for the unsorted case.

For lack of space we will in the following often only treat the (sufficient) subset $\{\wedge, \neg, !, \forall\}$ of logical symbols, since all others can be defined from these just as in the classical two-valued logic.

Kleene does not use the ! operator as a connective but treats it on the meta-level. While it is useful it is not necessary for the treatment. Furthermore, even this connective does not render $\mathcal{SKL}$ truth-functionally complete, since, just like the other connectives and the quantifiers, ! is *normal*, that is, when restricted to $\{\mathsf{f}, \mathsf{t}\}$ yields values in $\{\mathsf{f}, \mathsf{t}\}$.

Definition 2.7 (Σ-Model) Let A be a formula, then we call a strict Σ-algebra $\mathcal{M} := (\mathcal{A}, \mathcal{I})$ a *Σ-model for A* (written $\mathcal{M} \models A$), iff $\mathcal{I}_\varphi(A) = \mathsf{t}$ for all Σ-assignments φ. With this notion we can define the notions of ***validity***, ***(un)-satisfiability***, and ***entailment*** (i.e. $\Phi \models A$) in the usual way.

Remark 2.8 The "tertium non datur" principle of classical logic is no longer valid, since formulae can be undefined, in which case they are neither true nor false. We do, however, have a "quartum non datur" principle, that is, formulae are either true, false, or undefined, which allows us to derive the validity of a formula by refuting that it is false or undefined. We will use this observation in our tableau calculus.

The classical deduction theorem does not hold for $\mathcal{SKL}$ since the semantic status of a formula in the hypotheses is different from its status in the antecedent of an implication. A formula in the hypotheses is assumed to evaluate semantically to t, hence in particular it is defined. This leads to the following modified deduction theorem.

Theorem 2.9 (Deduction Theorem) *$\Phi \cup \{A\} \models B$ iff $\Phi \models A \wedge !A \Rightarrow B$.*

Proof: Let us first assume the first property and let $\mathcal{M}$ be a model of $\Phi \cup \{A\}$ then $\mathcal{M}$ is also a model of B, hence $\mathcal{M} \models A \wedge !A \Rightarrow B$. That means in order to show the second property we only have to look at interpretations which are models of Φ but not of A. For these, however, $A \wedge !A$ evaluates to f, hence they are models of $A \wedge !A \Rightarrow B$ too.

If the second property is given and $\mathcal{M}$ is a model of Φ then $\mathcal{M}$ is also a model of $A \wedge !A \Rightarrow B$. In order to prove the first property, only the subclass of those models has to be considered which are also models of A. These are, however, also models of $!A$, hence models of B too. □

Remark 2.10 While in classical logic, the consequence relation is directly connected to the implication, here things are a little bit more difficult. In particular, when proving mathematical theorems, it is quite usual to do this with respect to some background theory (axioms and definitions), which can no longer simply be taken in the antecedent of an implication. Hence we will often consider for mathematical applications so-called *consequents*, that is, pairs consisting of a set of formulae Φ and a formula A. We call a consequent $\Phi \models A$ valid, if A is entailed by Φ in all Σ-models.

Example 2.11 Now we can come back to the example from the exposition. The assertion is not a theorem of $\mathcal{SKL}$, since the instance $1 = \frac{1}{0} \Rightarrow 1 = 0 \cdot 1$ is not a valid formula (in any reasonable axiomatization of elementary arithmetic). While the antecedent of the implication evaluates to u, the succedent evaluates to f, hence the whole expression to u. Thus, this theorem cannot be derived in our sound tableau calculus to be presented in section 3.

Example 2.12 (Extended Example) We will formalize an extended example from elementary algebra that shows the basic features of $\mathcal{SKL}$. Here the sort $\mathbb{R}^*$ denotes the real numbers without zero. Note that we use the sort information to encode definedness information for inversion: $\frac{1}{x}$ is defined for all $x \in \mathbb{R}^*$, since $\mathbb{R}^*$ is subsort of $\mathfrak{D}$ by definition. Naturally, we give only a reduced formalization of real number arithmetic that is sufficient for our example. (For instance, we could add expressions like $\frac{1}{0}\not\Subset\mathfrak{D}$.) Consider the consequent $\{A1, A2, A3, A4, A5\} \models T$ with

A1 $\forall x_{\mathbb{R}}.\, x \neq 0 \Rightarrow x \Subset \mathbb{R}^*$
A2 $\forall x_{\mathbb{R}^*}.\, \frac{1}{x} \Subset \mathbb{R}^*$
A3 $\forall x_{\mathbb{R}^*}.\, x^2 > 0$
A4 $\forall x_{\mathbb{R}}.\, \forall y_{\mathbb{R}}.\, x - y \Subset \mathbb{R}$
A5 $\forall x_{\mathbb{R}}.\, \forall y_{\mathbb{R}}.\, x - y = 0 \Rightarrow x = y$
T $\forall x_{\mathbb{R}}.\, \forall y_{\mathbb{R}}.\, x \neq y \Rightarrow \left(\frac{1}{x-y}\right)^2 > 0$

An informal mathematical argumentation why T is entailed by $\{A1, \ldots, A5\}$ can be as follows: In the consequent above, the Ai are assumed to be true, that is, neither false nor undefined. Let x and y be arbitrary elements of $\mathbb{R}$. If $x = y$, the premise of T is false, hence the whole expression true (in this case the conclusion evaluates to u). If $x \neq y$, then the premise is true and the truth value of the whole expression is equal to that of the conclusion $\left(\frac{1}{x-y}\right)^2 > 0$. Since $x \neq y$ we get by A5 that $x - y \neq 0$ and by A4 that $x - y \Subset \mathbb{R}$, hence by A1 $x - y \Subset \mathbb{R}^*$ and by A2 $\frac{1}{x-y} \Subset \mathbb{R}^*$, which finally gives $\left(\frac{1}{x-y}\right)^2 > 0$ together with A3.

Note that this reasoning is not justified for the implication A := A1 $\wedge$ A2 $\wedge$ A3 $\wedge$ A4 $\wedge$ A5 $\Rightarrow$ T, since there are hidden assumptions, for instance, the totality of the binary predicate $>$ on $\mathbb{R} \times \mathbb{R}$. In fact the formula A is not a tautology, since it is possible to interpret the $>$ predicate as undefined for the second argument being zero, so that A3 as well as T evaluate to u, while the other Ai evaluate to t, hence the whole expression evaluates to u.

2.2 Relativization into Truth-Functional Logic

In this section we show that we can always systematically transform $\mathcal{SKL}$ formulae to formulae in an unsorted truth-functional three-valued logic $\mathbf{K}^3$ in a way that respects the semantics. However, we will see that this formulation will lose much of the conciseness of the presentation and enlarge the search spaces involved with automatic theorem proving.

At first glance it may seem that $\mathcal{SKL}$ is only a sorted variant of a three-valued instance of the truth functional many-valued logics that were very thoroughly investigated by Carnielli, Hähnle, Baaz and Fermüller [BF92, Car87, Car91, Häh92]. However, since all instances of this framework are truth-functional, that is, the denotations of the connectives and quantifiers only depend on the truth values of (certain instances of) their arguments, even unsorted Kleene logic does not fit into this paradigm, since quantification excludes the undefined element. In $\mathcal{SKL}$ we solve the problem with the quantification by postulating a sort $\mathfrak{D}$ of all defined individuals, which is a supersort of all other sorts. Therefore the relativization mapping not only considers sort information, it also has to care about definedness aspects in quantification.

Informally $\mathbf{K}^3$-formulae are just first-order formulae (with the additional unary connective !). While the three-valued semantics of the connectives is just that given in definition 2.6, the semantics of the quantifier uses unrestricted instantiation, that is,

$$\mathcal{I}_\varphi(\forall x.\, A) := \widetilde{\forall}(\{\mathcal{I}_{\varphi,[a/x]}(A) \mid a \in \mathcal{A}\})$$

Definition 2.13 (Relativization) We define transformations $\mathfrak{R}^{\mathcal{S}}$ and $\mathfrak{R}^{\mathfrak{D}}$, that map $\mathcal{SKL}$-sentences to unsorted $\mathcal{SKL}$-sentences and further into $\mathbf{K}^3$-sentences. $\mathfrak{R}^{\mathcal{S}}$ is the identity on terms and atoms, homomorphic on connectives, and

$$\mathfrak{R}^{\mathcal{S}}(\forall x_S.\, \Phi) := \forall x_{\mathfrak{D}}.\, S(x) \Rightarrow \mathfrak{R}^{\mathcal{S}}(\Phi)$$

Note that in order for these sentences to make sense in unsorted $\mathcal{SKL}$ we have to extend the set of predicate symbols by unary predicates S for all sort symbols $S \in \mathcal{S}^*$. Furthermore, for any of these new predicates we need the axiom: $\forall x_{\mathfrak{D}}.\, !S(x)$. The set of all these axioms is denoted by $\mathfrak{R}^{\mathcal{S}}(\Sigma)$.

We define $\mathfrak{R}^{\mathfrak{D}}$ to be the identity (only dropping the sort references from the variables) on terms and proper atoms and

- $\mathfrak{R}^{\mathfrak{D}}(t \lessdot \mathfrak{D}) := \mathfrak{D}(t)$
- $\mathfrak{R}^{\mathfrak{D}}(\forall x_{\mathfrak{D}}.\, A) := \forall x.\, \mathfrak{D}(x) \Rightarrow \mathfrak{R}^{\mathfrak{D}}(A)$

Just as above we have to extend the set of predicate symbols by a unary predicate $\mathfrak{D}$ and need a set $\mathfrak{R}^{\mathfrak{D}}(\Sigma)$ of signature axioms, which contains the axioms

$$\forall x_1, \ldots, x_n.\, P^n(x_1, \ldots, x_n) \vee \neg P^n(x_1, \ldots, x_n) \Rightarrow (\mathfrak{D}(x_1) \wedge \ldots \wedge \mathfrak{D}(x_n))$$
$$\forall x_1, \ldots, x_n.\, \mathfrak{D}(f(x_1, \ldots, x_n)) \Rightarrow (\mathfrak{D}(x_1) \wedge \ldots \wedge \mathfrak{D}(x_n))$$

for any predicate symbol $P \in \mathcal{P}^n$ and for any function symbol $f \in \mathcal{F}^n$, together with the axiom

$$\forall x.\, \mathfrak{D}(x) \vee \neg\mathfrak{D}(x)$$

These axioms axiomatize the $\mathcal{SKL}$ notion of definedness in $\mathbf{K}^3$. In particular the last axiom states that the predicate $\mathfrak{D}$ is two-valued, in contrast to all other sort predicates which are strict and thus three-valued. The other axioms force all functions and predicates to be interpreted strictly with respect to the $\mathfrak{D}$ predicate.

Theorem 2.14 (Sort Theorem) *Let Φ be a set of sentences, then the following statements are equivalent*

1. *Φ has a Σ-model.*
2. *$\Re^{\mathcal{S}}(\Phi)$ has a $\Sigma \cup \mathcal{S}^*$-model that satisfies $\Re^{\mathcal{S}}(\Sigma)$.*
3. *$\Re^{\mathfrak{D}} \circ \Re^{\mathcal{S}}(\Phi)$ has a $\mathbf{K}^3$-model that satisfies $\Re^{\mathfrak{D}}(\Sigma \cup \mathcal{S}^*) \cup \Re^{\mathfrak{D}}(\Re^{\mathcal{S}}(\Sigma))$.*

Proof: We will only show the equivalence of 2. and 3. since the equivalence of 1. and 2. can be proven with the same methods. Therefore we can restrict our proof to unsorted $\mathcal{SKL}$, where $\mathcal{S}^* = \emptyset$

Let $\mathcal{M} := (\mathcal{A}, \mathcal{I})$ be a Σ-model for Φ, then we construct a $\mathbf{K}^3$-model $\mathcal{M}^3 = (\mathcal{A}^3, \mathcal{I}^3)$ for $\Re^{\mathfrak{D}}(\Phi)$. Let $\mathcal{A}^3 := \mathcal{A}$, $\mathcal{I}^3(f) := \mathcal{I}(f)$ and $\mathcal{I}^3(P) := \mathcal{I}(P)$ where f is a function symbol and P is a predicate symbol or the sort $\mathfrak{D}$. Clearly, we have $\mathcal{M}^3 \models^{\mathbf{K}^3} \Re^{\mathfrak{D}}(\Sigma)$, since $\mathcal{M}$ is a Σ-model, where all functions are strict and the carrier $\mathcal{A}$, defined as the image of $\mathcal{I}^3(\mathfrak{D})$, is nonempty.

Furthermore let φ be a Σ-assignment and $\mathcal{M} \models_\varphi \Phi$, then we show by structural induction that $\mathcal{I}^3_\varphi(\Re^{\mathfrak{D}}(\Phi)) = \mathcal{I}_\varphi(\Phi)$ and therefore $\mathcal{M}^3 \models^{\mathbf{K}^3}_\varphi \Re^{\mathfrak{D}}(\Phi)$. This claim is immediate for terms and proper atoms. For sort atoms we have

$$\mathcal{I}^3_\varphi(\Re^{\mathfrak{D}}(t \lessdot \mathfrak{D})) = \mathcal{I}^3_\varphi(\mathfrak{D}(t)) = \mathcal{I}^3(\mathfrak{D})(\mathcal{I}^3_\varphi(t)) = \mathcal{I}(\mathfrak{D})(\mathcal{I}_\varphi(t)) = \mathcal{I}_\varphi(t \lessdot \mathfrak{D})$$

thus we have $\mathcal{I}^3_\varphi(\Re^{\mathfrak{D}}(A)) = \mathcal{I}_\varphi(A)$ for all atoms A. For quantified formulae we have

$$\mathcal{I}^3_\varphi(\Re^{\mathfrak{D}}(\forall x_{\mathfrak{D}}.\, \Psi)) = \mathcal{I}^3_\varphi(\forall x.\, \mathfrak{D}(x) \Rightarrow \Re^{\mathfrak{D}}(\Psi)) = \widetilde{\forall}(\Theta^3)\ ,$$

where $\Theta^3 := \{\mathcal{I}^3_\psi((\mathfrak{D}(x)) \Rightarrow \Re^{\mathfrak{D}}(\Psi)) \mid a \in \mathcal{A}^3\}$ and $\psi := \varphi, [a/x]$. On the other hand

$$\mathcal{I}_\varphi(\forall x_{\mathfrak{D}}.\, \Psi) = \widetilde{\forall}\{\mathcal{I}_\psi(\Psi) \mid a \in \mathcal{A}\} = \widetilde{\forall}(\Theta)$$

Now $\mathcal{I}^3_\varphi(\Re^{\mathfrak{D}}(\forall x_{\mathfrak{D}}.\, \Psi)) = \mathcal{I}^3_\varphi(\forall x.\, \mathfrak{D}(X) \Rightarrow \Re^{\mathfrak{D}}(\Psi))$
$= \widetilde{\forall}(\{\mathcal{I}^3_{\varphi,[a/x]}(\mathfrak{D}(X) \Rightarrow \Re^{\mathfrak{D}}(\Psi)) \mid a \in \mathcal{A}^3\}),$

so we have to consider the following cases for a. If $a = \bot$, then $\mathcal{I}^3_\psi(\mathfrak{D}(x)) = \mathsf{f}$ and therefore $\mathcal{I}^3_\psi(\mathfrak{D}(x) \Rightarrow \Re^{\mathfrak{D}}(\Psi)) = \mathsf{t}$. If $a \neq \bot$, then by inductive hypothesis $\mathcal{I}^3_\varphi(\Re^{\mathfrak{D}}(\Psi)) = \mathcal{I}_\varphi(\Psi)$ and therefore $\Theta^3 = \Theta \cup \{\mathsf{t}\}$.

$$\begin{array}{llll} \mathcal{I}^3_\varphi(\Re^{\mathfrak{D}}(\forall x_{\mathfrak{D}}.\, \Psi)) = \mathsf{t} & \text{iff} & \Theta^3 = \Theta = \{\mathsf{t}\} \text{ or } \emptyset & \text{iff} \quad \mathcal{I}_\varphi(\forall x_{\mathfrak{D}}.\, \Psi) = \mathsf{t} \\ \mathcal{I}^3_\varphi(\Re^{\mathfrak{D}}(\forall x_{\mathfrak{D}}.\, \Psi)) = \mathsf{u} & \text{iff} & \Theta^3 = \Theta = \{\mathsf{u},\mathsf{t}\} \text{ or } \{\mathsf{u}\} & \text{iff} \quad \mathcal{I}_\varphi(\forall x_{\mathfrak{D}}.\, \Psi) = \mathsf{u} \\ \mathcal{I}^3_\varphi(\Re^{\mathfrak{D}}(\forall x_{\mathfrak{D}}.\, \Psi)) = \mathsf{f} & \text{iff} & \mathsf{f} \in \Theta^3 = \Theta \cup \{\mathsf{t}\} & \text{iff} \quad \mathcal{I}_\varphi(\forall x_{\mathfrak{D}}.\, \Psi) = \mathsf{f} \end{array}$$

Since $\Re^{\mathfrak{D}}$ is homomorphic for connectives, we have completed the induction, thus $\mathcal{M}^3 \models^{\mathbf{K}^3} \Re^{\mathfrak{D}}(\Phi)$ and we have proven the necessitation direction of the theorem.

For the proof of sufficiency let $\mathcal{M}^3 := (\mathcal{A}^3, \mathcal{I}^3)$ be a $\mathbf{K}^3$-model, such that $\mathcal{M}^3 \models \Re^{\mathfrak{D}}(\Phi) \cup \Re^{\mathfrak{D}}(\Sigma)$, note that in our case $\Re^{\mathfrak{D}} \circ \Re^{\mathcal{S}} = \Re^{\mathfrak{D}}$. Let

$$\mathcal{A} := \{a \in \mathcal{A}^3 \mid \mathcal{I}^3(\mathfrak{D})(a) = \mathsf{t}\} \quad \text{and} \quad \mathcal{A}_\perp := \{a \in \mathcal{A}^3 \mid \mathcal{I}^3(\mathfrak{D})(a) = \mathsf{f}\}$$

then $\mathcal{A}^3 = \mathcal{A} \cup \mathcal{A}_\perp$, since $\forall x.\ \mathfrak{D}(x) \vee \neg\mathfrak{D}(x) \in \Re^{\mathfrak{D}}(\Sigma)$. If $\mathcal{A}_\perp = \emptyset$, then it is easy to construct a strict Σ-algebra from $(\mathcal{A}^3, \mathcal{I}^3)$ by extending $\mathcal{A}^3$ with $\perp$ and interpreting each function and predicate with the strict extension of its $\mathcal{I}^3$ value. So in the following we will assume that $\mathcal{A}_\perp$ is nonempty. Now let $\pi\colon \mathcal{A}^3 \longrightarrow \mathcal{A}^\perp$ be a function that is the identity on $\mathcal{A}$ and $\pi(a) = \perp$ for all $a \in \mathcal{A}_\perp$. As $\mathcal{M}^3 \models \Re^{\mathfrak{D}}(\Sigma)$, we know that $\mathcal{I}^3(f)(a_1, \ldots, a_n) \in \mathcal{A}_\perp$ if one $a_i \in \mathcal{A}_\perp$, so the following definition is well-defined.

$$\mathcal{I}(f)(\pi(a_1), \ldots, \pi(a_n)) := \pi(\mathcal{I}^3(f)(a_1, \ldots, a_n))$$

Now we will see that $\mathcal{I}_{\pi\circ\varphi}(t) = \pi(\mathcal{I}^3_\varphi(t))$ for all well-formed $\mathcal{SKL}$ terms t and assignments φ into $\mathcal{M}^3$.

1. $\mathcal{I}_{\pi\circ\varphi}(x) = \pi \circ \varphi(x) = \pi(\mathcal{I}^3_\varphi(x))$.
2. $\mathcal{I}_{\pi\circ\varphi}(c) = \mathcal{I}(c) = \pi(\mathcal{I}^3(c)) = \pi(\mathcal{I}^3_\varphi(c))$.
3. $\begin{aligned}[t] \mathcal{I}_{\pi\circ\varphi}(f(t^1, \ldots, t^n)) &= \mathcal{I}(f)(\mathcal{I}_{\pi\circ\varphi}(t^1), \ldots, \mathcal{I}_{\pi\circ\varphi}(t^n)) \\ &= \mathcal{I}(f)(\pi(\mathcal{I}^3_\varphi(t^1)), \ldots, \pi(\mathcal{I}^3_\varphi(t^n))) \\ &= \pi(\mathcal{I}^3(f)(\mathcal{I}^3_\varphi(t^1), \ldots, \mathcal{I}^3_\varphi(t^n))) \\ &= \pi(\mathcal{I}^3_\varphi(f(t^1, \ldots, t^n))) \end{aligned}$

Similarly the definition

$$\mathcal{I}(p)(\pi(a_1), \ldots, \pi(a_n)) := \mathcal{I}^3(p)(a_1, \ldots, a_n)$$

is well-defined, because $\mathcal{M}^3 \models \Re^{\mathfrak{D}}(\Sigma)$ and gives us $\mathcal{I}_{\pi\circ\varphi}(A) = \mathcal{I}^3(\Re^{\mathfrak{D}}(A))$ for all atoms A. From this, we obtain the general result $\mathcal{I}_{\pi\circ\varphi}(\Phi) = \mathcal{I}^3_\varphi(\Re^{\mathfrak{D}}(\Phi))$ by treating quantified formulae by a case analysis just as in the necessitation direction. In particular we have $\mathcal{I}_{\pi\circ\varphi}(\Phi) = \mathsf{t}$, iff $\mathcal{I}^3_\varphi(\Phi) = \mathsf{t}$ and therefore $\mathcal{M} \models \Phi$, whenever $\mathcal{M}^3 \models \Re^{\mathfrak{D}}(\Phi)$. □

Corollary 2.15 *Let Φ be a set of sentences and A be a sentence, then the following are equivalent*

1. *$\Phi \models A$ in all Σ-models.*
2. *$\Re^{\mathcal{S}}(\Phi) \cup \Re^{\mathcal{S}}(\Sigma) \models \Re^{\mathcal{S}}(A)$ in all unsorted $\Sigma \cup \mathcal{S}^*$-models.*
3. *$\Re^{\mathfrak{D}} \circ \Re^{\mathcal{S}}(\Phi) \cup \Re^{\mathfrak{D}}(\Sigma \cup \mathcal{S}^*) \cup \Re^{\mathfrak{D}}(\Re^{\mathcal{S}}(\Sigma)) \models \Re^{\mathfrak{D}} \circ \Re^{\mathcal{S}}(A)$ in all $\mathbf{K}^3$-models.*

As a consequence of the sort theorem, the standard operationalization for many-valued logics [BF92, Car87, Car91, Häh92] can be utilized to mechanize strong sorted Kleene logic. In fact, the system of Lucio-Carrasco and Gavilanes-Franco [LCGF89] can be seen as a standard many-valued tableau operationalization [Häh92, BFZ93] of the relativization of $\mathcal{SKL}$. However, as the extended example shows, we can do better by using sorted methods, since relativization expands the size and number of input formulae and furthermore expands the search spaces involved in automatic theorem proving by building up many meaningless branches. Note that already the formulation of $\mathcal{SKL}$ where we only have the required sort $\mathfrak{D}$ is much more concise than the relativized version. Furthermore we will see that the theory of definedness is treated goal-driven by the $\mathcal{TPF}$ calculus (cf. section 3). Thus the $\mathcal{TPF}$ calculus is closer to informal practice than the relativization in this respect.

Example 2.16 (continuing 2.12)
The relativization $\Re^{\mathcal{S}} \circ \Re^{\mathfrak{D}}$ of the $\mathcal{SKL}$-consequent $\{A1, A2, A3, A4, A5\} \models T$ is the $\mathbf{K}^3$-consequent $\{R1, R2, R3, R4, R5, R^{\mathbb{R}}, R^{\mathbb{R}^*}, R^{=}, R^{>}, R^{-}, R^{/}, R^{2}, \mathfrak{D}^{!}\} \models RT$ with the following relativized formulae:

R1 $\forall x.\, \mathfrak{D}(x) \Rightarrow (\mathbb{R}(x) \Rightarrow (x \neq 0 \Rightarrow \mathbb{R}^*(x)))$
R2 $\forall x.\, \mathfrak{D}(x) \Rightarrow (\mathbb{R}^*(x) \Rightarrow \mathbb{R}^*(\frac{1}{x}))$
R3 $\forall x.\, \mathfrak{D}(x) \Rightarrow (\mathbb{R}^*(x) \Rightarrow x^2 > 0)$
R4 $\forall x.\, \mathfrak{D}(x) \Rightarrow (\mathbb{R}(x) \Rightarrow (\forall y.\, \mathfrak{D}(y) \Rightarrow (\mathbb{R}(y) \Rightarrow \mathbb{R}(x-y))))$
R5 $\forall x.\, \mathfrak{D}(x) \Rightarrow (\mathbb{R}(x) \Rightarrow (\forall y.\, \mathfrak{D}(y) \Rightarrow (\mathbb{R}(y) \Rightarrow (x - y = 0 \Rightarrow x = y))))$
RT $\forall x.\, \mathfrak{D}(x) \Rightarrow (\mathbb{R}(x) \Rightarrow (\forall y.\, \mathfrak{D}(y) \Rightarrow (\mathbb{R}(y) \Rightarrow (x \neq y \Rightarrow \left(\frac{1}{x-y}\right)^2 > 0))))$

The set of signature axioms $\Re^{\mathfrak{D}}(\Sigma \cup \mathcal{S}^*) \cup \Re^{\mathfrak{D}}(\Re^{\mathcal{S}}(\Sigma))$ is the following set of $\mathbf{K}^3$-formulae:

$R^{\mathbb{R}}$ $\forall x.\, \mathfrak{D}(x) \Rightarrow !\mathbb{R}(x)$
$R^{\mathbb{R}^*}$ $\forall x.\, \mathfrak{D}(x) \Rightarrow !\mathbb{R}^*(x)$
$R^{=}$ $\forall x, y.\, (x = y \vee x \neq y) \Rightarrow \mathfrak{D}(x) \wedge \mathfrak{D}(y)$
$R^{>}$ $\forall x, y.\, (x > y \vee x \not> y) \Rightarrow \mathfrak{D}(x) \wedge \mathfrak{D}(y)$
R^{-} $\forall x, y.\, \mathfrak{D}(x - y) \Rightarrow \mathfrak{D}(x) \wedge \mathfrak{D}(y)$
$R^{/}$ $\forall x.\, \mathfrak{D}(\frac{1}{x}) \Rightarrow \mathfrak{D}(x)$
R^{2} $\forall x.\, \mathfrak{D}(x^2) \Rightarrow \mathfrak{D}(x)$
$\mathfrak{D}^{!}$ $\forall x.\, \mathfrak{D}(x) \vee \neg\mathfrak{D}(x)$

2.3 Model Existence

In this subsection we introduce an important tool for proving the completeness of calculi. The importance of model existence theorems lies in the fact that they abstract over the model theoretic part of various completeness proofs. Such theorems were first introduced by Smullyan (who calls them unifying principles) in [Smu63, Smu68] based on work by Hintikka and Beth.

Definition 2.17 Let ∇ be a class of sets.

1. ∇ is called *closed under subsets*, iff for all sets S and T the following condition holds: if $S \subset T$ and $T \in \nabla$, then $S \in \nabla$.
2. ∇ is called *compact*, iff for every set S of formulae: $S \in \nabla$, iff all finite subsets of S are members of ∇ too.

Lemma 2.18 *If ∇ is compact, then ∇ is closed under subsets.*

Proof: Suppose $S \subset T$ and $T \in \nabla$. Every finite subset A of S is a finite subset of T, and since ∇ is compact, we know that $A \in \nabla$. Thus $S \in \nabla$. □

Definition 2.19 (Labeled Formula) We will call a pair A^α, where A is an $\mathcal{SKL}$-formula and $\alpha \in \{\mathsf{f}, \mathsf{u}, \mathsf{t}\}$ a *labeled formula.* We say that a Σ-assignment φ satisfies a set Φ of labeled formulae in a strict Σ-algebra $\mathcal{M} = (\mathcal{A}, \mathcal{I})$, if $\mathcal{I}_\varphi(A) = \alpha$ for all $A^\alpha \in \Phi$.

In the following we will use $\Phi * A$ as an abbreviation for $\Phi \cup \{A\}$ in order to increase the legibility.

Definition 2.20 (Abstract Consistency Class) A class ∇ of sets of labeled formulae is called an *abstract consistency class*, iff it is closed under subsets, and for all sets $\Phi \in \nabla$ the following conditions hold:

1. If A is atomic, then $A^\alpha \in \Phi$ for at most one $\alpha \in \{\mathsf{f}, \mathsf{u}, \mathsf{t}\}$, furthermore for all terms t the literal $(t \lessdot \mathfrak{D})^\mathsf{u}$ is not in Φ.
2. If $(\neg A)^\alpha \in \Phi$, then $\Phi * A^\beta \in \nabla$, where $\beta = \mathsf{t}$, if $\alpha = \mathsf{f}$; $\beta = \mathsf{f}$, if $\alpha = \mathsf{t}$; and $\beta = \mathsf{u}$ else.
3. If $(!A)^\mathsf{t} \in \Phi$, then $\Phi * A^\gamma \in \nabla$ for some $\gamma \in \{\mathsf{t}, \mathsf{f}\}$; if $(!A)^\mathsf{f} \in \Phi$, then $\Phi * A^\mathsf{u} \in \nabla$. $(!A)^\mathsf{u}$ is not in any $\Phi \in \nabla$.
4. If $(A \vee B)^\alpha \in \Phi$, then
   $\alpha = \mathsf{t}$) $\Phi * A^\mathsf{t} \in \nabla$ or $\Phi * B^\mathsf{t} \in \nabla$.
   $\alpha = \mathsf{u}$) $\Phi \cup \{A^\mathsf{f}, B^\mathsf{u}\} \in \nabla$, or $\Phi \cup \{A^\mathsf{u}, B^\mathsf{f}\} \in \nabla$, or $\Phi \cup \{A^\mathsf{u}, B^\mathsf{u}\} \in \nabla$.
   $\alpha = \mathsf{f}$) $\Phi \cup \{A^\mathsf{f}, B^\mathsf{f}\} \in \nabla$
5. If $\forall x_S. A \in \Phi$, then
   $\alpha = \mathsf{t}$) for any term t, $\Phi * ([t/x_S]A)^\mathsf{t} \in \nabla$ or $\Phi * (t \lessdot S)^\alpha \in \nabla$ for some $\alpha \in \{\mathsf{f}, \mathsf{u}\}$.
   $\alpha = \mathsf{u}$) for any term t, and any constant c that does not occur, $\Phi \cup \{([c/x_S]A)^\mathsf{u}, (c \lessdot S)^\mathsf{t}, \mathcal{A}\} \in \nabla$, where $\mathcal{A}$ is $([t/x_S]A)^\mathsf{t}$ or $([t/x_S]A)^\mathsf{u}$ or $(t \lessdot S)^\mathsf{f}$ or $(t \lessdot S)^\mathsf{u}$.
   $\alpha = \mathsf{f}$) $\Phi \cup \{([c/x]A)^\mathsf{f}, (c \lessdot S)^\mathsf{t}\} \in \nabla$, for each constant c that does not occur in Φ.
6. If $A^\gamma \in \Phi$ with $\gamma \in \{\mathsf{f}, \mathsf{t}\}$, then $\Phi * (t \lessdot \mathfrak{D})^\mathsf{t} \in \nabla$, for all subterms t of A.
7. If $(t \lessdot S)^\mathsf{u} \in \Phi$, then $\Phi * (t \lessdot \mathfrak{D})^\mathsf{f} \in \nabla$.

Theorem 2.21
For each abstract consistency class ∇ there exists an abstract consistency class ∇' such that $\nabla \subset \nabla'$, and ∇' is compact.

Proof: (following [And86]) Let $\nabla' := \{\Phi \mid \text{every finite subset of } \Phi \text{ is in } \nabla\}$. To see that $\nabla \subset \nabla'$, suppose that $\Phi \in \nabla$. ∇ is closed under subsets, so every finite subset of Φ is in ∇, and thus $\Phi \in \nabla'$.

Next let us show that ∇' is compact. Suppose $\Phi \in \nabla'$ and Ψ is an arbitrary finite subset of Φ. By definition of ∇' all finite subsets of Ψ are in ∇, and therefore $\Psi \in \nabla'$. Thus all finite subsets of Φ are in ∇' whenever Ψ is in ∇'. On the other hand, suppose all finite subsets of Ψ are in ∇'. Then by the definition of ∇' the finite subsets of Ψ are also in ∇, so $\Phi \in \nabla'$. Thus ∇' is compact.

Finally we show that ∇' is an abstract consistency class. By lemma 2.18 it is closed under subsets. Of the conditions for the abstract consistency class we will only explicitly present the first two cases, since the proofs of the others are analogous. Let $\Phi \in \nabla'$ be given arbitrarily.

Suppose there is an atom A, such that $\{A^\alpha, A^\beta\} \subseteq \Phi$ for $\alpha \neq \beta$. By the definition of ∇' we get $\{A^\alpha, A^\beta\} \in \nabla$ contradicting 2.20(1).

Let $(\neg A)^\alpha \in \Phi$, and Ψ be any finite subset of $\Phi * A^\beta$ (where α and β are as in 2.20(2)) and let $\Theta := (\Psi \setminus \{A^\beta\}) * (\neg A)^\alpha$. Θ is a finite subset of Φ, so $\Theta \in \nabla$. Since ∇ is an abstract consistency class and $(\neg A)^\alpha \in \Theta$, we get $\Theta * A^\beta \in \nabla$. We know that $\Psi \subset \Theta * A^\beta$, and ∇ is closed under subsets, so $\Psi \in \nabla$. Thus every finite subset Ψ of $\Phi * A^\beta$ is in ∇, therefore by definition $\Phi * A^\beta \in \nabla'$. □

Definition 2.22 (Σ-Hintikka Set) Let ∇ be an abstract consistency class and $\Phi \in \nabla$. Then $\mathcal{H} \in \nabla$ is called a *∇-extension of Φ*, iff $\Phi \subset \mathcal{H}$. A set $\mathcal{H}$ is called *maximal in* ∇, iff for each formula $D \in \nabla$ such that $\mathcal{H} * D \in \nabla$, we already have $D \in \mathcal{H}$. A set $\mathcal{H} \in \nabla$ is called a *Σ-Hintikka set for ∇ and Φ*, iff $\mathcal{H}$ is maximal in ∇ and $\Phi \subseteq \mathcal{H}$.

We now give some technical properties of Σ-Hintikka sets that are useful for manipulating formulae.

Theorem 2.23 *If ∇ is an abstract consistency class, and $\mathcal{H}$ is maximal in ∇, then the following statements hold:*

1. *If A is a proposition, then $A^\alpha \in \mathcal{H}$ for at most one $\alpha \in \{\mathsf{f}, \mathsf{u}, \mathsf{t}\}$. Furthermore $(t \lessdot \mathfrak{D})^{\mathsf{u}} \notin \mathcal{H}$ for all t.*
2. *If $(\neg A)^\alpha \in \mathcal{H}$, then $A^\beta \in \mathcal{H}$, where $\beta = \mathsf{t}$, if $\alpha = \mathsf{f}$; $\beta = \mathsf{f}$, if $\alpha = \mathsf{t}$; and $\beta = \mathsf{u}$ else.*
3. *If $(!A)^\alpha \in \mathcal{H}$, then either $\alpha = \mathsf{t}$ and $A^\gamma \in \mathcal{H}$ for $\gamma \in \{\mathsf{f}, \mathsf{t}\}$ or $\alpha = \mathsf{f}$ and $A^{\mathsf{u}} \in \mathcal{H}$. In particular, there is no formula B, such that $(!B)^{\mathsf{u}} \in \mathcal{H}$.*
4. *If $(A \vee B)^\alpha \in \mathcal{H}$, then*
   $\alpha = \mathsf{t}$) *$A^{\mathsf{t}} \in \mathcal{H}$ or $B^{\mathsf{t}} \in \mathcal{H}$.*
   $\alpha = \mathsf{u}$) *$A^{\mathsf{f}}, B^{\mathsf{u}} \in \mathcal{H}$, or $A^{\mathsf{u}}, B^{\mathsf{f}} \in \mathcal{H}$, or $A^{\mathsf{u}}, B^{\mathsf{u}} \in \mathcal{H}$.*
   $\alpha = \mathsf{f}$) *$A^{\mathsf{f}}, B^{\mathsf{f}} \in \mathcal{H}$*
5. *If $\forall x_S.\, A \in \mathcal{H}$, then*
   $\alpha = \mathsf{t}$) *for any term t, $[t/x_S]A^{\mathsf{t}} \in \mathcal{H}$ or $(t \lessdot S)^\alpha \in \mathcal{H}$ for some $\alpha \in \{\mathsf{f}, \mathsf{u}\}$.*
   $\alpha = \mathsf{u}$) *for any term t, there is a term s, with $\Phi \cup \{([s/x_S]A)^{\mathsf{u}}, (s \lessdot S)^{\mathsf{t}}, \mathcal{A}\} \in \mathcal{H}$, where $\mathcal{A} \in \{([t/x_S]A)^{\mathsf{t}}, ([t/x_S]A)^{\mathsf{u}}, (t \lessdot S)^{\mathsf{f}}\}$.*
   $\alpha = \mathsf{f}$) *there is a term t, such that $([t/x]A)^{\mathsf{f}}, (t \lessdot S)^{\mathsf{t}} \in \mathcal{H}$.*
6. *If $A^\gamma \in \mathcal{H}$ with $\gamma \in \{\mathsf{t}, \mathsf{f}\}$, then $(t \lessdot \mathfrak{D})^{\mathsf{t}} \in \mathcal{H}$, for all subterms t of A.*
7. *If $(t \lessdot S)^{\mathsf{u}} \in \mathcal{H}$, then $(t \lessdot \mathfrak{D})^{\mathsf{f}} \in \mathcal{H}$.*

Proof: We prove the first assertion by induction on the structure of A. If A is atomic, then the assertion is a simple consequence of 2.20(1).

Let $A = \neg B$ and $A^{\mathsf{f}}, A^{\mathsf{t}}$ be in $\mathcal{H}$. By 2.20(2) we have $B^{\mathsf{f}}, B^{\mathsf{t}} \in \mathcal{H}$ contradicting the induction hypothesis. The remaining cases can be shown analogously, so we have proven the first assertion.

The rest of the assertions are all of the same form, and have analogous proofs, therefore we only prove the second. If $(\neg A)^{\mathsf{f}} \in \mathcal{H}$, then $\mathcal{H} * A^{\mathsf{t}} \in \nabla$ (∇ is an abstract consistency class). The maximality of $\mathcal{H}$ now gives the assertion. □

Lemma 2.24 (Hintikka Lemma) *If ∇ is an abstract consistency class and $\mathcal{H}$ is maximal in ∇, then there is an $\mathcal{SKL}$-model $\mathcal{M}$ and a Σ-assignment φ, such that φ satisfies $\mathcal{H}$ in $\mathcal{M}$.*

Proof: We prove the assertion by constructing a model $\mathcal{M} = (\mathcal{A}, \mathcal{I})$ for $\mathcal{H}$, which is derived from the ground term algebra.

Let $\mathcal{T}_\perp$ be the set of closed well-formed terms together with $\perp$. In order to construct the carrier $\mathcal{A}$, we have to identify all elements in $\mathcal{T}_\perp$ that are undefined $((t{<}\mathfrak{D})^{\mathsf{f}} \in \mathcal{H})$ and identify them with $\perp$. Traditional proofs of the Hintikka-Lemma for total-function logics now define $\mathcal{I}\colon \mathcal{F} \longrightarrow \mathcal{A}$ to be the identity map. However, this definition does not make $\mathcal{I}(f)$ strict, since $\mathcal{I}(f)(\perp) = f(\perp) \neq \perp$. To repair this defect we take the carrier $\mathcal{A}$ to be the quotient of $\mathcal{T}_\perp$ with respect to the equality theory $=_\perp$ induced by the set

$$E_\perp = \{t =_\perp \perp \mid (t{<}\mathfrak{D})^{\mathsf{f}} \in \mathcal{H}\} \cup \{f^k(x_1, \ldots, \perp, \ldots, x_k) = \perp \mid f^k \in \Sigma^k\}$$

of equations. Thus $\mathcal{A}$ is the set of equivalence classes $[\![t]\!]_\perp = \{s \mid E_\perp \models s =_\perp t\}$. The function $f_\perp\colon ([\![t_1]\!]_\perp, \ldots, [\![t_1]\!]_\perp) \mapsto [\![f(t_1, \ldots, t_n)]\!]_\perp$ is a well-defined function, since $=_\perp$ is a congruence relation. We define $\mathcal{I}(f) := f_\perp$ and note that the special construction of $E_\perp$ entails the strictness of $f_\perp$.

For any finite set $\mathcal{W}$ of variables a Σ-assignment φ can be restricted to a substitution $\varphi_\mathcal{W} = \varphi|_\mathcal{W}$. A simple induction on the structure of a term t can be used to show that $\mathcal{I}_\varphi(t) = \mathcal{I}_{\varphi_{\mathbf{Free}(t)}}(t)$.

For $P \in \mathcal{P}^n$ let $\mathcal{I}(P)\colon \mathcal{A}^n \longrightarrow \{\mathsf{f}, \mathsf{u}, \mathsf{t}\}$ with $P_\mathcal{H}([\![t^1]\!]_\perp, \ldots, [\![t^n]\!]_\perp) = \alpha$, iff $P(t^1, \ldots, t^n)^\alpha \in \mathcal{H}$. Clearly $P_\mathcal{H}$ is a partial function (cf. 2.23.1), since the definition only depends on $=_\perp$-equivalence classes. With the help of 2.23.6 it is easy to see that $P_\mathcal{H}$ is a strict function. We can extend $P_\mathcal{H}$ to a total strict function $\mathcal{I}(P)$ by evaluating all remaining proper atoms with u and all remaining sort atoms with f. Thus sorts are everywhere defined (the value u is only obtained on $\perp$) and the strictness of the predicates is preserved.

Clearly, this construction entails that for any atom $A \in \mathcal{H}$ and any Σ-assignment φ we have $\mathcal{I}_\varphi(A) = \alpha$, iff $\mathcal{I}_{\varphi_{\mathbf{Free}(A)}}(A) \in \mathcal{H}$. Now a simple induction on the number of connectives and quantifications, using the properties of 2.23 can be used to extend this property to arbitrary formulae. Thus we have $\mathcal{I}_\varphi(A) = \alpha$ for all $A^\alpha \in \mathcal{H}$, if we take φ to be the identity. □

We now come to the proof of the abstract extension lemma, which nearly immediately yields the model existence theorem.

Theorem 2.25 (Abstract Extension Lemma) *Let ∇ be a compact abstract consistency class, and let $H \in \nabla$ be a set of propositions. Then there exists a Σ-Hintikka set $\mathcal{H}$ for ∇ and H.*

Proof: We will construct $\mathcal{H}$ by inductively constructing a sequence of sets $\mathcal{H}^i$ and taking $\mathcal{H} := \bigcup_{i \in \mathbb{N}} \mathcal{H}^i \in \nabla$. We can arrange all labeled formulae in an infinite sequence $C^1, C^2, \ldots$ For each $n \in \mathbb{N}$ we inductively define a set $\mathcal{H}^n$ of propositions by

1. $\mathcal{H}^0 := \Phi$.
2. If $\mathcal{H}^n * C^n \notin \nabla$, then $\mathcal{H}^{n+1} := \mathcal{H}^n$.
3. If $\mathcal{H}^n * C^n \in \nabla$, and C^n is of the form $(\forall x_S.\ A)^\alpha$ with $\alpha \in \{\mathsf{f}, \mathsf{u}\}$ then $\mathcal{H}^{n+1} := \mathcal{H}^n \cup \{C^n, [c^n/x]A^\alpha, (c \lessdot S)^{\mathsf{t}}\}$.
4. $\mathcal{H}^{n+1} := \mathcal{H}^n * C^n$ else.

Let $\mathcal{H} := \bigcup_{n \in \mathbb{N}} \mathcal{H}^n$. Clearly each of the $\mathcal{H}^n \in \nabla$, and therefore $\mathcal{H} \in \nabla$, since ∇ is compact.

In order to prove the maximality of $\mathcal{H}$, let A be an arbitrary proposition such that $\mathcal{H} * A \in \nabla$. We know that $A = C^n$ for some $n \in \mathbb{N}$, so $\mathcal{H}^n * A \subset \mathcal{H} * A \in \nabla$ and $\mathcal{H}^n * A \in \nabla$, since ∇ is closed under subsets. Hence by definition we know that $A \in \mathcal{H}^{n+1}$, and therefore $A \in \mathcal{H}$. □

Corollary 2.26 (Model Existence) *Let $\Phi \in \nabla$ and ∇ be an abstract consistency class, then there is an $\mathcal{SKL}$-model $\mathcal{M}$ and a Σ-assignment φ, such that φ satisfies Φ in $\mathcal{M}$.*

Proof: Let ∇' be the compact abstract consistency class of theorem 2.21 and let $\mathcal{H}$ be the maximal ∇-extension of Φ guaranteed by 2.25. Furthermore let $\mathcal{M}$ be the $\mathcal{SKL}$-model, and φ the Σ-assignment for $\mathcal{H}$ guaranteed by 2.24. Then φ satisfies Φ in $\mathcal{M}$, since $\Phi \subseteq \mathcal{H}$. □

3 Tableau

Now we turn to the exposition of our tableau calculus. The case of standard tableaux for partial functions is a simple extension of first-order tableau methods to $\mathcal{SKL}$. Therefore we will only concern ourselves with free variable tableaux.

While a labeled formula A^α means that A has the truth value α, we also make use of multi-indices as introduced by Hähnle and write $A^{\alpha\beta}$ as an abbreviation for $A^\alpha \vee A^\beta$. (Normally, we do not have to consider three different truth values, since the corresponding formulae are tautological and cannot contribute to refutations.) As has been pointed out by Hähnle [Häh92], the use of multi-indices does not only offer a concise notation, but can drastically improve a calculus, when special rules for their treatment are introduced. In the following, we add corresponding rules for handling multi-indices, where one label is u. Although not necessary in principle, this treatment results in a significant improvement of the search complexity of the calculus, which can thereby be reduced to the complexity in the two-valued case. This relationship will be made formal in theorem 3.9.

Definition 3.1 (Tableau Rules) The tableau rules consist of the traditional tableau rules for the propositional connectives, augmented by the case of the label u.

$$\frac{(A\wedge B)^{\mathsf{t}}}{\begin{array}{c}A^{\mathsf{t}}\\ B^{\mathsf{t}}\end{array}} \qquad \frac{(A\wedge B)^{\mathsf{u}}}{\begin{array}{c}A^{\mathsf{ut}}\\ B^{\mathsf{ut}}\\ A^{\mathsf{u}}\mid B^{\mathsf{u}}\end{array}} \qquad \frac{(A\wedge B)^{\mathsf{f}}}{A^{\mathsf{f}}\mid B^{\mathsf{f}}} \qquad \frac{(A\wedge B)^{\mathsf{ut}}}{\begin{array}{c}A^{\mathsf{ut}}\\ B^{\mathsf{ut}}\end{array}} \qquad \frac{(A\wedge B)^{\mathsf{fu}}}{A^{\mathsf{fu}}\mid B^{\mathsf{fu}}}$$

Since we have special rules for the multi-indices ut and fu, we only need a splitting rule reflecting the definition of multi-indices as disjunctions for the remaining multi-index ft. Note that the multi-index fut gives rise to tautologies, which can never contribute to refutations.

$$\frac{A^{\mathsf{ft}}}{A^{\mathsf{f}}\mid A^{\mathsf{t}}}$$

The negation rules just flip the labels in the intuitive way.

$$\frac{(\neg A)^{\mathsf{t}}}{A^{\mathsf{f}}} \qquad \frac{(\neg A)^{\mathsf{u}}}{A^{\mathsf{u}}} \qquad \frac{(\neg A)^{\mathsf{f}}}{A^{\mathsf{t}}} \qquad \frac{(\neg A)^{\mathsf{ut}}}{A^{\mathsf{fu}}} \qquad \frac{(\neg A)^{\mathsf{fu}}}{A^{\mathsf{ut}}}$$

The ! rule for the u case closes the branch (we use an explicit symbol $*$ for that), since $(!A)^{\mathsf{u}}$ is unsatisfiable in $\mathcal{SKL}$.

$$\frac{(!A)^{\mathsf{t}}}{A^{\mathsf{ft}}} \qquad \frac{(!A)^{\mathsf{u}}}{*} \qquad \frac{(!A)^{\mathsf{f}}}{A^{\mathsf{u}}} \qquad \frac{(!A)^{\mathsf{ut}}}{A^{\mathsf{ft}}} \qquad \frac{(!A)^{\mathsf{fu}}}{A^{\mathsf{u}}}$$

In order to simplify the presentation of the examples we also (redundantly) present the rules for disjunction.

$$\frac{(A\vee B)^{\mathsf{t}}}{A^{\mathsf{t}}\mid B^{\mathsf{t}}} \qquad \frac{(A\vee B)^{\mathsf{u}}}{\begin{array}{c}A^{\mathsf{fu}}\\ B^{\mathsf{fu}}\\ A^{\mathsf{u}}\mid B^{\mathsf{u}}\end{array}} \qquad \frac{(A\vee B)^{\mathsf{f}}}{\begin{array}{c}A^{\mathsf{f}}\\ B^{\mathsf{f}}\end{array}} \qquad \frac{(A\vee B)^{\mathsf{ut}}}{A^{\mathsf{ut}}\mid B^{\mathsf{ut}}} \qquad \frac{(A\vee B)^{\mathsf{fu}}}{\begin{array}{c}A^{\mathsf{fu}}\\ B^{\mathsf{fu}}\end{array}}$$

The quantifier rules for the classical truth values and multi-indices are very similar to the standard rules[2] ($\{x_S, y^1, \ldots, y^n\}$ are the free variables of A and f is a new function symbol of arity n), with the exception that the sort of the Skolem function has to be specified. The rule for the case u has a mixed existential and universal character: for y_S the value of A is undefined or true

[2] We employ the liberalized δ-rule of [HS94].

(that is there is no instance, which makes the formula false) *and* there is at least one witness for the undefinedness.

$$\frac{(\forall x_S.\, A)^{\mathsf{t}}}{[y_S/x_S]A^{\mathsf{t}}} \qquad \frac{(\forall x_S.\, A)^{\mathsf{u}}}{\begin{array}{c}[f(y^1,\ldots,y^n)/x_S]A^{\mathsf{u}}\\ [y_S/x_S]A^{\mathsf{ut}}\\ (f(y^1,\ldots,y^n){<}S)^{\mathsf{t}}\end{array}} \qquad \frac{(\forall x_S.\, A)^{\mathsf{f}}}{\begin{array}{c}[f(y^1,\ldots,y^n)/x_S]A^{\mathsf{f}}\\ (f(y^1,\ldots,y^n){<}S)^{\mathsf{t}}\end{array}}$$

$$\frac{(\forall x_S.\, A)^{\mathsf{ut}}}{[y_S/x_S]A^{\mathsf{ut}}} \qquad \frac{(\forall x_S.\, A)^{\mathsf{fu}}}{\begin{array}{c}[f(y^1,\ldots,y^n)/x_S]A^{\mathsf{fu}}\\ (f(y^1,\ldots,y^n){<}S)^{\mathsf{t}}\end{array}}$$

The rules for connectives and quantifiers above can now be used to reduce complex labeled formulae to literals. Some sort literals can further be reduced, due to the fact that sorts are defined on all defined individuals and the predicate $\mathfrak{D}$ is defined everywhere. (These rules have to be slightly generalized for multi-indices. We only display the interesting case.)

$$\frac{(t{<}\mathfrak{D})^{\mathsf{u}}}{*} \qquad \frac{(t{<}S)^{\mathsf{u}}}{(t{<}\mathfrak{D})^{\mathsf{f}}}$$

Now we only need tableau closure rules: The *cut* rule and the *strict* rule

$$\frac{\begin{array}{c}A^{\alpha}\\ B^{\beta}\end{array}}{* \mid \mathcal{SC}(\sigma)}\,\sigma \qquad \frac{\begin{array}{c}C^{\gamma}\\ (t{<}\mathfrak{D})^{\mathsf{f}}\end{array}}{* \mid \mathcal{SC}(\sigma)}\,\sigma$$

where $\alpha \cap \beta = \emptyset$, $\gamma \subseteq \{\mathsf{ft}\}$, and $\sigma = [t_1/x^1_{S_1}], \ldots, [t_n/x^n_{S_n}]$ is the most general unifier of A and B or the most general unifier of the term t and a subterm s of C, respectively. In both cases the *sort constraint* $\mathcal{SC}(\sigma) = ((t_1{<}S_1)\wedge\ldots\wedge(t_n{<}S_n))^{\mathsf{fu}}$ insures the correctness (in terms of the sorts) of the instantiations. We have employed the notation of writing the substitution σ next to the tableau schema, to indicate that the whole tableau is instantiated by σ during the application of the rule.

A tableau is built up by constructing a tree with the tableau rules starting with an initial tree without branchings. We call a tableau *closed*, iff all of its branches end in $*$. Note that the disjunct $*$ in the succedent of the rules above is only needed, if the set of sort constraints is empty. Then this rule closes the branch without residuating.

Remark 3.2 We could also have used a generalization of the cut rule of the form

$$\frac{\begin{array}{c}A^{\alpha}\\ B^{\beta}\end{array}}{A^{\alpha\cap\beta} \mid \mathcal{SC}(\sigma)}\,\sigma$$

where we employ the convention that $A^{\emptyset} = *$, since this corresponds to the empty disjunction, which is unsatisfiable. However, it is not straightforward to see in which cases this variant of the cut rule is more efficient.

Definition 3.3 (Tableau Proof) A *tableau proof for a formula* A is a closed tableau constructed from the initial tree consisting of the labelled formula A^{fu}. A *tableau proof for a consequent* $\Phi \models A$ is a closed tableau constructed from $\Phi^{\mathsf{t}} \cup \{A^{\mathsf{fu}}\}$.

Remark 3.4 The tableau proof of a consequent $\Phi \models A$ essentially refutes the possibility that A can be undefined or false under the assumption of all formulae in Φ. By the quartum non datur rule, we can then conclude that A is entailed by Φ.

3.1 Soundness and Completeness

The soundness of the $\mathcal{TPF}$ rules can be verified by a tedious recourse to the semantics of the quantifiers and connectives. Completeness is proven by the standard argument using the model existence theorem for $\mathcal{SKL}$. For this, we first have to prove a lifting theorem for $\mathcal{TPF}$

Theorem 3.5 (Tableau Lifting) *Let $\Phi \models A$ be a consequent and θ a substitution, then $\Phi \models A$ has a closed $\mathcal{TPF}$-tableau provided $\theta(\Phi) \models \theta(A)$ has one.*

Proof: Let $\mathcal{T}_\theta$ be a closed tableau for $\theta(\Phi) \models \theta(A)$, the claim is proven by an induction on the construction of $\mathcal{T}_\theta$ constructing a tableau $\mathcal{T}$ for $\Phi \models A$ that is tableau-isomorphic to $\mathcal{T}$. Concretely we have a tree-isomorphism $\omega: \mathcal{T} \longrightarrow \mathcal{T}_\theta$ between $\mathcal{T}_\theta$ and $\mathcal{T}$ that respects labels and is compatible with θ, that is, for any node $\mathcal{N}$ in $\mathcal{T}$ with labeled formula A^α we have $\omega_{\mathcal{N}}(A) = \theta(A)$.

This induction is straightforward for all $\mathcal{TPF}$ rules except for the cut and the strict rules that residuate a sort constraint. In both cases, we can use a standard argument which we will only carry out for the cut case: σ is a most general unifier of $\theta(A)$ and $\theta(B)$ in $\mathcal{T}_\theta$, so $\sigma \circ \theta$ unifies A and B in $\mathcal{T}$. So there exists a most general unifier ρ of A and B, and a substitution τ with $\sigma \circ \theta = \tau \circ \rho$. Now we have $\tau(\mathcal{SC}(\rho)) = \mathcal{SC}(\tau \circ \rho) = \mathcal{SC}(\sigma \circ \theta)$, so we obtain the assertion by the inductive hypothesis for $\rho(\mathcal{T})$ and $\tau \circ \rho(\mathcal{T}) = \sigma(\mathcal{T}_\theta)$. □

Theorem 3.6 (Completeness) *$\mathcal{TPF}$ is refutation complete, that is, if $\Phi \models A$ is a valid consequent, then there is a closed tableau for $\Phi^{\mathsf{t}} \cup A^{\mathsf{fu}}$.*

Proof: Completeness of $\mathcal{TPF}$ can be proven using the model existence theorem 2.26 by verifying that the class ∇ of sets Φ that do not have closed $\mathcal{TPF}$ tableaux is an abstract consistency class. This can be achieved with the usual techniques of e.g. [Fit90]: It is obvious that the $\mathcal{TPF}$ rules for the connectives, quantifiers and sorts directly correspond the clauses of 2.20. We have treated the only case where this correspondence is nontrivial (the quantifier case) in the tableau lifting theorem above. □

Example 3.7 (continuing 2.12) Taking the above example we give a proof for

$$\{A1, A2, A3, A4, A5\} \models T$$

using the above tableau rules. The proof is shown in figure 1. Applying the closure rule in the case of non-empty sort constraints, we omit the $*$ branch for simplicity reasons. Note that the unsorted unifiers $[c - d/u_{\mathrm{IR}}]$, $[c/x_{\mathrm{IR}}]$, and $[d/y_{\mathrm{IR}}]$ have to be applied to the whole tableau. For display reasons, however, we only add the relevant formulae to the tableau instead of replacing them, that is, correctly (F8) has to replace (F3) and (F13) to replace (F9).

The tableau proof can roughly be divided into three different parts, first the representation of the problem, displayed above the first line, second some initial simplification by eliminating quantifiers and connectives displayed between the first and the second line, and third the final refutation, below the second line.

Remark 3.8 The proof in figure 1 shows an interesting feature, namely it corresponds in length and structure exactly to a proof of the theorem in classical two-valued logic. By replacing all truth-value sets fu by the truth value f you get the corresponding two-valued proof. This correspondence is due to the correspondence of the tableau rules R^{α} and $R^{\alpha\mathsf{u}}$ for $\alpha \in \{\mathsf{f}, \mathsf{t}\}$ and $R \in \{\wedge, \vee, \neg, \forall\}$. In other words using rules for truth-value sets provides proofs as short as in the two-valued case. If, however, truth-value sets are not used, certain parts of the proofs must be duplicated. This relationship can only hold for so-called *normal problems* of course, that is, problems which do not contain any ! connective, since formulae containing a ! do not make any sense in classical two-valued logic.

Theorem 3.9 (Correspondence Theorem) *Each tableau proof for a normal problem $\Phi \models A$ in $\mathcal{SKL}$ can be isomorphically transformed into a tableau proof in $\mathcal{FOL}$.*

Proof: Let us prove the assumption by a case analysis on the rules applied in the proof. At a certain formula in the $\mathcal{SKL}$-tableau, its label set either contains the u value or not. If the formula does not contain u then it is labeled by t, by f, or by ft and will be treated by the same rule R^{α} with $R \in \{\wedge, \vee, \neg, \forall\}$ and $\alpha \in \{\mathsf{f}, \mathsf{t}\}$ or the splitting rule. Note the corresponding tableau rules are the same for $\mathcal{FOL}$ and $\mathcal{SKL}$.

In the case that α contains the truth value u, just eliminate u from the set. Since the initial problem formulation contains only the labels t and fu, for normal problems it inductively follows that no formula with the label u can occur in a tableau. The procedure of just eliminating the truth value u is correct, since for all connectives (with the exception of !, which may not occur in normal problems), all quantifier and all truth values we can verify that if R is a rule in $\mathcal{SKL}$ with a truth value set containing the truth value u, then a tableau rule of $\mathcal{FOL}$ can be constructed from R by eliminating the truth value u in the rule. For instance

$$\frac{(A \wedge B)^{\mathsf{ut}}}{\begin{array}{c} A^{\mathsf{ut}} \\ B^{\mathsf{ut}} \end{array}} \quad \leadsto \quad \frac{(A \wedge B)^{\mathsf{t}}}{\begin{array}{c} A^{\mathsf{t}} \\ B^{\mathsf{t}} \end{array}}$$

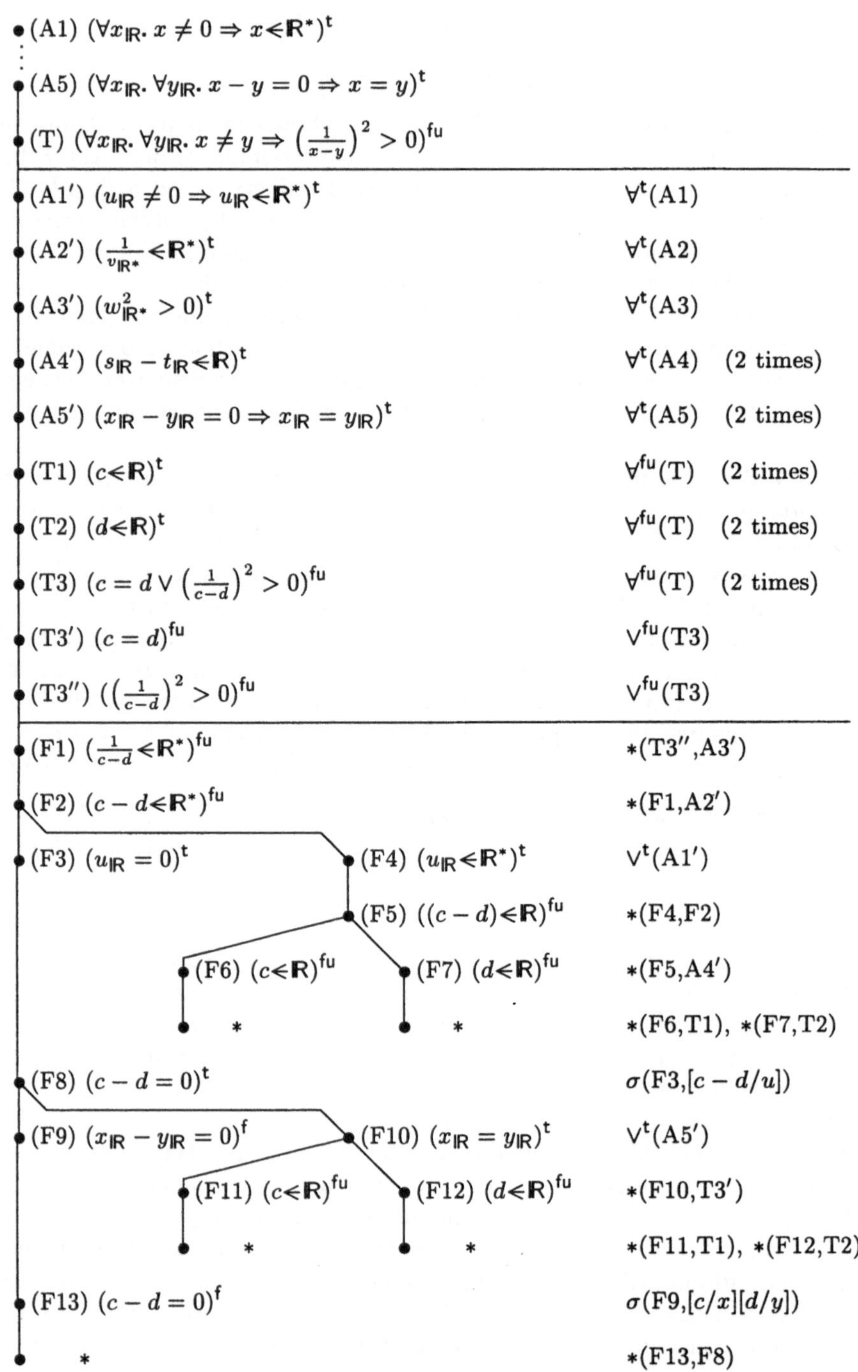

(A1) $(\forall x_{\mathbb{R}}.\, x \neq 0 \Rightarrow x \lessdot \mathbb{R}^*)^{\mathsf{t}}$
(A5) $(\forall x_{\mathbb{R}}.\, \forall y_{\mathbb{R}}.\, x - y = 0 \Rightarrow x = y)^{\mathsf{t}}$
(T) $(\forall x_{\mathbb{R}}.\, \forall y_{\mathbb{R}}.\, x \neq y \Rightarrow (\frac{1}{x-y})^2 > 0)^{\mathsf{fu}}$
(A1′) $(u_{\mathbb{R}} \neq 0 \Rightarrow u_{\mathbb{R}} \lessdot \mathbb{R}^*)^{\mathsf{t}}$ $\forall^{\mathsf{t}}$(A1)
(A2′) $(\frac{1}{v_{\mathbb{R}^*}} \lessdot \mathbb{R}^*)^{\mathsf{t}}$ $\forall^{\mathsf{t}}$(A2)
(A3′) $(w_{\mathbb{R}^*}^2 > 0)^{\mathsf{t}}$ $\forall^{\mathsf{t}}$(A3)
(A4′) $(s_{\mathbb{R}} - t_{\mathbb{R}} \lessdot \mathbb{R})^{\mathsf{t}}$ $\forall^{\mathsf{t}}$(A4) (2 times)
(A5′) $(x_{\mathbb{R}} - y_{\mathbb{R}} = 0 \Rightarrow x_{\mathbb{R}} = y_{\mathbb{R}})^{\mathsf{t}}$ $\forall^{\mathsf{t}}$(A5) (2 times)
(T1) $(c \lessdot \mathbb{R})^{\mathsf{t}}$ $\forall^{\mathsf{fu}}$(T) (2 times)
(T2) $(d \lessdot \mathbb{R})^{\mathsf{t}}$ $\forall^{\mathsf{fu}}$(T) (2 times)
(T3) $(c = d \vee (\frac{1}{c-d})^2 > 0)^{\mathsf{fu}}$ $\forall^{\mathsf{fu}}$(T) (2 times)
(T3′) $(c = d)^{\mathsf{fu}}$ $\vee^{\mathsf{fu}}$(T3)
(T3″) $((\frac{1}{c-d})^2 > 0)^{\mathsf{fu}}$ $\vee^{\mathsf{fu}}$(T3)
(F1) $(\frac{1}{c-d} \lessdot \mathbb{R}^*)^{\mathsf{fu}}$ $*$(T3″,A3′)
(F2) $(c - d \lessdot \mathbb{R}^*)^{\mathsf{fu}}$ $*$(F1,A2′)
(F3) $(u_{\mathbb{R}} = 0)^{\mathsf{t}}$ (F4) $(u_{\mathbb{R}} \lessdot \mathbb{R}^*)^{\mathsf{t}}$ $\vee^{\mathsf{t}}$(A1′)
(F5) $((c - d) \lessdot \mathbb{R})^{\mathsf{fu}}$ $*$(F4,F2)
(F6) $(c \lessdot \mathbb{R})^{\mathsf{fu}}$ (F7) $(d \lessdot \mathbb{R})^{\mathsf{fu}}$ $*$(F5,A4′)
$*$ $*$ $*$(F6,T1), $*$(F7,T2)
(F8) $(c - d = 0)^{\mathsf{t}}$ σ(F3,[$c - d/u$])
(F9) $(x_{\mathbb{R}} - y_{\mathbb{R}} = 0)^{\mathsf{f}}$ (F10) $(x_{\mathbb{R}} = y_{\mathbb{R}})^{\mathsf{t}}$ $\vee^{\mathsf{t}}$(A5′)
(F11) $(c \lessdot \mathbb{R})^{\mathsf{fu}}$ (F12) $(d \lessdot \mathbb{R})^{\mathsf{fu}}$ $*$(F10,T3′)
$*$ $*$ $*$(F11,T1), $*$(F12,T2)
(F13) $(c - d = 0)^{\mathsf{f}}$ σ(F9,[c/x][d/y])
$*$ $*$(F13,F8)

Fig. 1. Tableau proof with unsorted unification, example 3.7

For the other cases check this relation in definition 3.1. This relation holds also for the tableau closure rule.

Thus we get a $\mathcal{FOL}$ proof from the $\mathcal{SKL}$ proof by simply eliminating all truth values u. □

Remarks 3.10 Unfortunately, the converse of the above theorem does not hold. Not each $\mathcal{FOL}$ proof can be transformed into an $\mathcal{SKL}$ proof, even if there is an $\mathcal{SKL}$ proof. Consider for example the relation $\{A\} \models A \vee (B \vee \neg B)$ which holds in $\mathcal{SKL}$ as well as in $\mathcal{FOL}$. An $\mathcal{FOL}$-proof is:

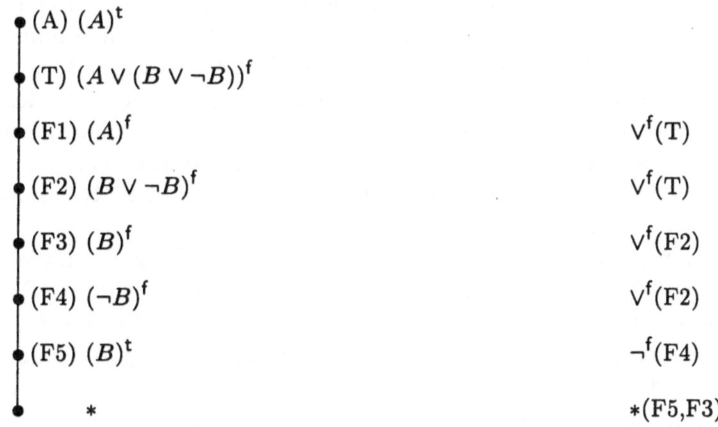

Fig. 2. Counterexample to the converse correspondence theorem

This proof cannot be transferred since in $\mathcal{SKL}$ (T), (F1), (F2), (F3), (F4), and (F5) are labeled by the truth value u in addition, hence the closure rule does not apply to (F5) and (F3). This comes from the fact that $B \vee \neg B$ is not a tautology in $\mathcal{SKL}$. However, the other straightforward closure of the tableau by applying the closure rule to (A) and (F1) can be applied in $\mathcal{FOL}$ as well as in $\mathcal{SKL}$.

Of course it would be nice to have the property that for each classical $\mathcal{FOL}$ proof there exists an $\mathcal{SKL}$ proof which is as short as the classical (of course only if the classical theorem is also an $\mathcal{SKL}$ theorem). The example above shows that this property does not hold in general, for instance, replace the assumption set $\{A\}$ by a set from which A can be derived in 20 steps only. On the other hand this example is rather artificial insofar as the theorem would normally not be stated in this form in mathematics, because mathematical theorems are normally not redundant in the way that two true statements are linked by an "$\vee$", on the contrary *usual* mathematical theorems employ preconditions as weak as possible and consequences as strong as possible. For instance, in a mathematical context we would expect theorems like A, $B \vee \neg B$, $A \wedge (B \vee \neg B)$. While a proof for the first (from the assumptions A) can be transferred from $\mathcal{FOL}$ to $\mathcal{SKL}$, the latter two are not theorems in $\mathcal{SKL}$. Hence we expect that for usual mathematical theorems the proof effort in $\mathcal{SKL}$ will not be bigger then in $\mathcal{FOL}$.

4 Extensions – Sorted Unification

Even though the $\mathcal{TPF}$ calculus defined above represents a significant computational improvement over a naive tableau calculus for Kleene's strong logic for partial functions, it only makes very limited use of the sorts in $\mathcal{SKL}$. This can be improved by utilizing a rigid sorted unification algorithm that takes into account all the sort information present in the respective branch and uses it as a local sort signature. This measure in effect restricts the set of possible unifiers to those that are well-sorted with respect to this (local) sort signature. This allows to perform some of the reasoning about well-sortedness (and therefore definedness) in the unification in an algorithmic way. This reasoning would otherwise be triggered by the sort constraints in $\mathcal{SKL}$ and would have to be carried out in the proof search. The methods presented in this section are heavily influenced by Weidenbach's work on sorted tableau methods in [Wei94].

In the tableau framework the extension with sorted unification is simpler (but perhaps less powerful) than in the resolution framework (see for instance [Wei91, KK93]). The reason for this is the difference in the treatment of the disjunction in resolution and tableau. Tableau calculi use the β rule to analyze disjuncts in different branches, but pay the price with the necessity to instantiate the entire tableau. Consider, for example, the formula $t \lessdot S \vee t \lessdot T$ stating that the term t has sort S or sort T. In the tableau method, we can investigate both situations in separate branches (with different local sets of declarations). In the resolution method, we have to use one of the disjuncts for sorted unification and residuate the other as a constraint, which has to be attached to well-sorted terms, well-sorted substitutions and clauses resulting from resolutions, whenever the other literal is used. On the other hand, the tableau method needs to instantiate all declarations that are used, since they can contain variables that also appear in other branches. Consider, for example, the axiom $\forall x_{\mathbb{R}}.\, x > 0 \Rightarrow x \lessdot \mathbb{R}^*$, which can be read as a conditional declaration. This axiom will result in branches containing the literal $(x_{\mathbb{R}} > 0)^{\mathsf{f}}$ and the declaration $(x_{\mathbb{R}} \lessdot \mathbb{R}^*)^{\mathsf{t}}$. If we use the declaration in sorted unification to justify that $1 \lessdot \mathbb{R}^*$, then we have to refute that $(1 > 0)^{\mathsf{f}}$ in the other branch. This simple example shows that we have to use (not surprisingly in a tableau framework) a *rigid* variant of sorted unification for our extension.

Definition 4.1 (Rigid Sorted Unification) Let $\mathcal{D}$ be a set of declarations, then we call a substitution σ *rigidly well-sorted* with respect to $\mathcal{D}$, iff there is a substitution τ, such that

1. $\sigma \subseteq \tau$ and $\mathbf{Dom}(\tau) \subseteq \mathbf{Free}(\mathcal{D}) \cup \mathbf{Dom}(\sigma)$
2. τ is well-sorted with respect to $\tau(\mathcal{D})$.

For instance the substitution $\sigma = [f(f(x_S))/z_S]$ is well-sorted, but not rigidly so, for the set $\mathcal{D} = \{f(y_S){<}S\}$, since the declaration has to be used twice (in differing instances) to show that $f(f(x_S))$ has sort S. σ is, however, rigidly well-sorted with respect to $\mathcal{D}' = \{f(y_S){<}S, f(v_S){<}S\}$, and the substitution $\tau = [f(f(x_S))/z_S], f(v_s)/y_s]$ is a substitution that instantiates $\mathcal{D}'$ in the appropriate way.

4.1 Rigid Sorted Unification

Sorted unification with term declarations was first considered by Schmidt-Schauß who also presents a sound and complete algorithm in [SS89]. In $\mathcal{SKL}$, term declarations appear as sort atoms of the form $t{<}S$, declaring all instances of t to be of sort S. Rigid sorted unification is treated in [Wei94].

Definition 4.2 (Well-Sorted Terms) Let $\mathcal{D}$ be a set of *declarations* (positive sort literals of the form $(t{<}S)^{\mathsf{t}}$), then the set $\mathbf{wsT}_S(\mathcal{D})$ of *well-sorted terms of sort S* is inductively defined by

1. variables $x_S \in \mathbf{wsT}_S(\mathcal{D})$
2. if $t{<}T \in \mathcal{D}$ then $t \in \mathbf{wsT}_T(\mathcal{D})$
3. if $t \in \mathbf{wsT}_T(\mathcal{D})$ and $s \in \mathbf{wsT}_S(\mathcal{D})$ then $[s/x_S]t \in \mathbf{wsT}_T(\mathcal{D})$.

We call a substitution $[t^1/x^1_{S_1}], \ldots, [t^1/x^n_{S_n}]$ a *well-sorted substitution*, iff $t^i \in \mathbf{wsT}_{S_i}(\mathcal{D})$. Obviously the application of well-sorted substitutions to well-sorted terms yields well-sorted terms, so $\mathbf{wsT}(\mathcal{D})$ is closed under well-sorted substitutions and the set of well-sorted substitutions is a monoid with function composition.

Remark 4.3 The definition above is an inductive one, not in the structure of terms, but in the justification of the well-sortedness. A simple induction on this justification shows that the consequent $\mathcal{D} \models t{<}S$ is valid for any term $t \in \mathbf{wsT}_S(\mathcal{D})$. In particular, for any well-sorted term $t \in \mathbf{wsT}_S(\mathcal{D})$ the denotation $\mathcal{I}_\varphi(t)$ is in $\mathcal{A}_S$ and therefore defined.

Furthermore a declaration of the form $x_S{<}T \in \mathcal{D}$ entails that $\mathbf{wsT}_S(\mathcal{D}) \subseteq \mathbf{wsT}_T(\mathcal{D})$ and $\mathcal{A}_S \subseteq \mathcal{A}_T$ for any Σ-model $\mathcal{A}$ of $\mathcal{D}$. Therefore we call declarations of the form $x_S{<}T \in \mathcal{D}$ *subsort declarations.*

Since we are working in a tableau framework and our sorted unification algorithm involves nondeterminism, we utilize the tableau search mechanism for the search for unifiers by representing unification constraints as special dis-equality literals. This gives us a very uniform presentation of the combined tableau procedure

Definition 4.4 (Tableau Rules for Rigid Sorted Unification)
We assume the existence of a binary predicate symbol $\doteq \in \mathcal{P}^2$ and call a literal $(s \doteq t)^{\mathsf{fu}}$ a *constraint literal* and often abbreviate the $(s \doteq t)^{\mathsf{fu}}$ by $s \not\doteq t$, as usual we do not distinguish between $(s \doteq t)$ and $(t \doteq s)$. We model sorted unification

as a tableau-based constraint simplification calculus with the following set of inference rules: The *decomposition* rule

$$\frac{f(s_1,\ldots,s^n) \not\doteq f(t^1,\ldots,t^n)}{* \mid s^1 \not\doteq t^1 \mid \ldots \mid s^n \not\doteq t^n}$$

is just the traditional decomposition transformation, known from unification theory. Again note that we only need the disjunct $*$, if $n = 0$. The *subsort* rule

$$\frac{\begin{array}{c}(z_T \lessdot S)^{\mathfrak{t}}\\ x_S \not\doteq y_T\end{array}}{*}\ [z_T/x_S],[z_T/y_T]$$

allows to eliminate variables, provided that T is a subsort of S, in which case the instantiation $[z_T/x_S]$ is well-sorted. The *intersect* rule

$$\frac{\begin{array}{c}(u_V \lessdot S)^{\mathfrak{t}}\\ (v_V \lessdot T)^{\mathfrak{t}}\\ x_S \not\doteq y_T\end{array}}{*}\ [u_V/x_S],[u_V/y_T],[u_V/v_V]$$

allows to eliminate a pair of variables that share a common subsort V. Finally a pair of variables can be eliminated for a term t, if t has sorts S and T, even if they do not share a common subsort (we call this situation *irregular*). Therefore the following rule *non-reg* instantiates the variables with the least committed generalization of t.

$$\frac{\begin{array}{c}(f(s_1,\ldots,s_n) \lessdot S)^{\mathfrak{t}}\\ (f(t_1,\ldots,t_n) \lessdot T)^{\mathfrak{t}}\\ x_S \not\doteq y_T\end{array}}{* \mid s^1 \not\doteq t^1 \mid \ldots \mid s^n \not\doteq t^n}\ [f(s^1,\ldots,s^n)/x_S],[f(t^1,\ldots,t^n)/y_T]$$

Finally we need a rule (the *imitation rule* below) that allows to eliminate a variable x_S for a term t, if it is an instance of a declaration in the branch above.

$$\frac{\begin{array}{c}(f(t^1,\ldots,t^n) \lessdot S)^{\mathfrak{t}}\\ x_S \not\doteq f(s^1,\ldots,s^n)\end{array}}{* \mid s^1 \not\doteq t^1 \mid \ldots \mid s^n \not\doteq t^n}\ [f(t^1,\ldots,t^n)/x_S]$$

In contrast to the related set of rules for sorted unification in [Wei91] or [SS89] we only eliminate solved pairs that are known to be well-sorted from the set $\mathcal{D}$ of declarations. Therefore we do not need the explicit failure rules these authors need, since they do not test for well-sortedness of the pair before eliminating. In our system we define failure as irreducibility and non-solvedness, but we could also add explicit failure rules to detect failure early for a practical implementation.

We say that a declaration $(t \lessdot S)^{\mathfrak{t}}$ is *used* by a unification inference rule, if it appears in the antecedent of the rule.

Theorem 4.5 *The above set of rules define a sound and complete non-determinist unification algorithm.*

Proof sketch: It is obvious that all inference rules maintain the property of well-sortedness for unification problems, since all new pairs added are from declarations and are therefore well-sorted by definition and the set of well-sorted terms is closed under well-sorted substitutions. Since the set of inference rules is a rigid variant of that given in [SS89, p.98], we refer to the proofs given there. These only have to be modified to account for rigidity. A close inspection of the differences shows that Schmidt-Schauß's rules can be obtained from ours by renaming all declarations that are used by the unification rules before applying the rules, and thus preventing that the declarations are used up in the process. For the proof of completeness, we construct a rigid extension τ from a non-rigid unifier by taking into account the instantiations of the declarations (in the rigid set of rules) that were circumvented in Schmidt-Schauß's rules by renaming. □

4.2 A Tableau Calculus for $\mathcal{SKL}$ using Rigid Sorted Unification

We will now extend $\mathcal{TPF}$ with a variant of the rigid sorted unification alogrithm above. Note that the notion of substitution discussed above is still not appropriate for a refutation calculus, where substitutions need to have ground instances. Otherwise the tableau cut rule becomes unsound: Let S be a sort that does not have ground terms, that is, where $\mathcal{A}_S$ may be empty, then a branch containing the literals $(Px_S)^{\mathrm{t}}$ and $(Py_S)^{\mathrm{f}}$ could be closed using the substitution $[y_S/x_S]$, without being unsatisfiable. A well-sorted term may not have ground instances, if it contains variables of sorts that do not have ground terms. Therefore we are interested in conditions for sorts to be non-empty.

Lemma 4.6 *Let $\mathcal{D}$ be a set of sort declarations, then the problem whether the set of ground terms of sort S is empty is decidable.*

Proof sketch: Let $Ax(\mathcal{D})$ be the set of propositional formulae $(S^1 \wedge \ldots \wedge S^n) \Rightarrow T$, such that $t \lessdot T \in \mathcal{D}$ and $\{x^1_{S^i}\}$ are the free variables of t. Then the emptiness problem is equivalent to the problem whether $Ax(\mathcal{D}) \models S$ in propositional logic, which is known to be decidable. □

Remark 4.7 Thus we can modify the sorted unification algorithm above by allowing tableau closure $*$ (or equivalently the rule to be applicable) only iff the sorts of the free variables in the substitutions associated with the rules are non-empty with respect to the set $\mathcal{D}$ of declarations in the branch above. This variant of the sorted unification algorithm only returns sorted unifiers that have well-sorted ground instances.

Now we will present an extension $\mathcal{TPF}(\Sigma)$ of $\mathcal{TPF}$ that allows to restrict the calculation to formulae that are well-sorted with respect to the declarations present on the branch above, and thereby prune branches of the proof search that would not lead to refutations, since they contain meaningless objects.

Definition 4.8 (Tableau with Sorted Unification ($\mathcal{TPF}(\Sigma)$))
To obtain the tableau calculus $\mathcal{TPF}(\Sigma)$ with sorted unification from $\mathcal{TPF}$, we modify the tableau closure rules and add the modified (cf. 4.7) constraint variant of the constraint simplification rules of sorted unification. The new *cut* and *strict* rules have the form

$$\frac{\begin{array}{c} A^{\alpha} \\ B^{\beta} \end{array}}{A \neq B} \qquad \frac{\begin{array}{c} C^{\gamma} \\ (t \lessdot \mathfrak{D})^{\mathsf{f}} \end{array}}{s \neq t}$$

where $\alpha \cap \beta = \emptyset$, C has a subterm s, and $\gamma \subseteq \{\mathsf{f},\mathsf{t}\}$. Thus instead of using unsorted unification, these rules residuate a unification constraint that can then be processed by the sorted unification algorithm. All other $\mathcal{TPF}$ rules stay unchanged.

Theorem 4.9 *$\mathcal{TPF}(\Sigma)$ is sound and refutation complete.*

Proof sketch: The soundness of $\mathcal{TPF}(\Sigma)$ relies on the soundness of the sorted unification algorithm, which guarantees only well-sorted instantiations. For the completeness proof we can again use the model existence theorem 2.26, where we only have to reconsider the tableau lifting theorem for $\mathcal{TPF}(\Sigma)$. This can be proven with the standard argumentation, since the sorted unification algorithm is complete and we can abstract from the internal structure of the unification derivation. □

As we have seen in remark 4.3 the well-sorted substitutions and therefore well-sorted unifications filter out instantiations of the tableau that contain meaningless objects and therefore cannot contribute to a refutation of the initial consequent. This property yields a significant pruning of the search spaces and therefore in a gain of computational efficiency. However, the rigidity of the unification algorithm makes it necessary to guess in advance the number of instances of the declarations needed for a proof, since they are used up during the unification. This is especially bothersome, since in general a great multiplicity of declarations is needed. In order to arrive at a more practical algorithm it will be important to find variants of the unification algorithm that are rigid only on the disjunctive part of the declarations present in a consequent.

Example 4.10 (continuing 3.7) Now we revisit the problem of proving

$$\{\mathrm{A1}, \mathrm{A2}, \mathrm{A3}, \mathrm{A4}, \mathrm{A5}\} \models \mathrm{T}$$

using the tableau calculus with sorted unification. While the first two main parts of the proof in figure 1, namely the problem setting and the initial simplification remain the same, the proper refutation will be shorter, in particular only three branches instead of five have to be considered. In figure 3, we display only the last part.

The unification for closing F2 with T3′ is straightforward because of T1 and T2, $(c \lessdot \mathbb{R})^{\mathsf{t}}$ and $(d \lessdot \mathbb{R})^{\mathsf{t}}$, while the sorted unification for closing F3 and F5 makes use of T1, T2, and the term declaration A4′. For the closure of T3″ and A3′ the unification algorithm must derive that $\frac{1}{c-d}$ has the sort $\mathbb{R}^*$, this is done by F6 with the term declaration A2′.

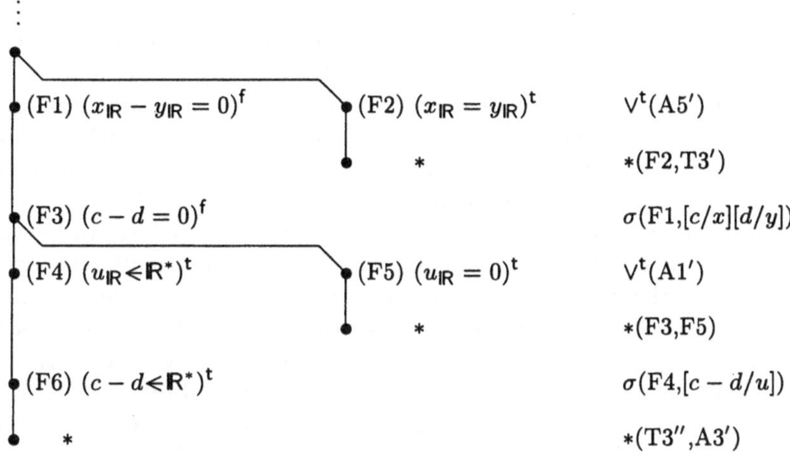

Fig. 3. Tableau proof with sorted unification

5 Conclusion

We have developed a sorted three-valued logic for the formalization of informal mathematical reasoning with partial functions. This system generalizes the system proposed by Kleene in [Kle52] for the treatment of partial functions over natural numbers to general first-order logic. In fact we believe that the unsorted version of our system without the ! operator is a faithful formalization of Kleene's ideas.

If we compare $\mathcal{SKL}$ to the three other approaches mentioned in the introduction, we see that the truth conditions coincide on valid mathematical statements, but that $\mathcal{SKL}$ properly excludes statements that a mathematician would reject as having problems with definedness. While the first approach has not the necessary expressiveness, the second and fourth approaches legitimate unwanted statements as theorems.

We have presented a sound and complete tableau calculus with dynamic sorts for our logic $\mathcal{SKL}$, which uses the sort mechanism to capture the fact that in Kleene's logic quantification only ranges over defined individuals. Our calculus can be seen as an extension of classical logic that combines methods from many-valued logics (cf. [BF92, Häh92]) for a correct treatment of the undefined and sorted logics (see [Wei89, Wei91]) for an adequate treatment of the defined. It differs from the sequent calculus in [LCGF89] in that the use of dynamic sort techniques greatly simplifies the calculus, since most definedness preconditions can be taken care of in the unification. Thus we believe that our system is not only more faithful to Kleene's ideas (definedness inference is handled in the unification at a level below the calculus) but also more efficient for the sort techniques involved.

In an earlier work [KK93, KK94] we had represented a resolution calculus of strong sorted Kleene logic. In this work, we not only have transferred the methods developed there to the tableau framework, but have also shown that normally proofs that keep track of the definedness conditions are not more complex than those in the classical two-valued logic. In some sense it is surprising that in spite of the advantages mentioned above, the complexity of proof search can be preserved by the treatment of multi-indices.

Of course further extensions of the system described here have to be considered in order to be feasible for practical mathematics. In particular this calculus does not address the question of the mechanization of higher-order features for the formalization of mathematical practice. Higher-order logics are especially suitable for formalizing partial functions, since functions are first class objects of the systems, that can even be quantified over. In this direction the work of Farmer et al. [Far90, FGT93] has shown that partial functions are a very natural and powerful tool for formalizing mathematics. We expect that our three-valued approach, which remedies some problems of their simpler two-valued approach (see the discussion in the introduction and in example 2.11) can be generalized in much the same manner and will be a useful tool for formalizing mathematics.

Finally, the authors believe that the merit of the idea of generalizing first-order logic with respect to both, the number of truth values and the domain of quantification is not confined to the application to partial functions. In particular there seem to be no obstacles against the extension of many multi-valued logics in artificial intelligence (such as Belnap's four-valued paraconsistent logic) that have only been investigated for the propositional fragment to the first-order case using our techniques.

References

[And86] Peter B. Andrews. *An Introduction to Mathematical Logic and Type Theory: To Truth through Proof.* Academic Press, 1986.

[Bee85] Michael J. Beeson. *Foundations of Constructive Mathematics.* Springer Verlag, 1985.

[Bet55] E. W. Beth. Semantic entailment and formal derivability. *Medelingen von de Koninklijke Nederlandse Akademie van Wetenschappen, Afdeling Letterkunde*, 18(13):309–342, 1955.

[BF92] Matthias Baaz and Christian G. Fermüller. Resolution for many-valued logics. In A. Voronkov, editor, *Proceedings of LPAR*, pages 107–118, St. Petersburg, Russia, 1992. Springer Verlag, LNAI 624.

[BFZ93] Matthias Baaz, Christian G. Fermüller, and Richard Zach. Dual systems of sequents and tableaux for many-valued logics. Technical Report TUW-E185.2BFZ.2-92, Technische Universität Wien, 1993.

[Car87] Walter A. Carnielli. Systematization of finite many-valued logics through the method of tableaux. *The Journal of Symbolic Logic*, 52:473–493, 1987.

[Car91] Walter A. Carnielli. On sequents and tableaux for many-valued logics. *Journal of Non-Classical Logic*, 8(1):59–76, 1991.

[Far90] William M. Farmer. A partial functions version of Church's simple theory of types. *The Journal of Symbolic Logic*, 55(3):1269–1291, 1990.

[FGT93] William M. Farmer, Joshua D. Guttman, and F. Javier Thayer. IMPS: An Interactive Mathematical Proof System. *Journal of Automated Reasoning*, 11(2):213–248, October 1993.

[Fit90] Melvin Fitting. *First-Order Logic and Automated Theorem Proving*. Springer Verlag, 1990.

[Häh92] Reiner Hähnle. *Automated Theorem Proving in Multiple Valued Logics*. PhD thesis, Fachbereich Informatik, Universität Karlsruhe, Karlsruhe, Germany, March 1992. revised version: Automated Deduction in Multiple-Valued Logics, Oxford University Press, 1994.

[Hin55] K. J. J. Hintikka. Form and content in quantification theory. *Acta Philosophica Fennica*, 8:7–55, 1955.

[HS94] Reiner Hähnle and Peter H. Schmitt. The liberalized δ-rule in free variable tableaux. *Journal of Automated Reasoning*, 12(2):211–222, 1994.

[KK93] Manfred Kerber and Michael Kohlhase. A mechanization of strong Kleene logic for partial functions. SEKI Report SR-93-20, Fachbereich Informatik, Universität des Saarlandes, Im Stadtwald, Saarbrücken, Germany, 1993.

[KK94] Manfred Kerber and Michael Kohlhase. A mechanization of strong Kleene logic for partial functions. In Alan Bundy, editor, *Proceedings of the 12th CADE*, pages 371–385, Nancy, France, 1994. Springer Verlag, LNAI 814.

[Kle52] Stephen C. Kleene. *Introduction to Metamathematics*. Van Nostrand, 1952.

[LCGF89] Francisca Lucio-Carrrasco and Antonio Gavilanes-Franco. A first order logic for partial functions. In *Proceedings STACS'89*, pages 47–58. Springer Verlag, LNCS 349, 1989.

[Pra60] Dag Prawitz. An improved proof procedure. *Theoria*, 26:102–139, 1960.

[Ree87] S. Reeves. Semantic tableaux as a framework for automated theorem-proving. In J. Hallam and C. Mellish, editors, *Advances in Artificial Intelligence, AISB-87*, pages 125–139. Wiley, 1987.

[Sch68] R. Schock. *Logics without Existence Assumptions*. Almquist & Wisell, 1968.

[Sco70] Dana S. Scott. Outline of a mathematical theory of computation. Technical Monograph PRG-2, Oxford University Computing Laboratory, 1970.

[Smu63] Raymond M. Smullyan. A unifying principle for quantification theory. *Proc. Nat. Acad Sciences*, 49:828–832, 1963.

[Smu68] Raymond M. Smullyan. *First-Order Logic*. Springer Verlag, 1968.

[SS89] Manfred Schmidt-Schauß. *Computational Aspects of an Order-Sorted Logic with Term Declarations*, Springer Verlag, LNAI 395, 1989.

[Tic82] Pawel Tichy. Foundations of partial type theory. *Reports on Mathematical Logic*, 14:59–72, 1982.

[Wei89] Christoph Weidenbach. A resolution calculus with dynamic sort structures and partial functions. SEKI Report SR-89-23, Fachbereich Informatik, Universität Kaiserslautern, Kaiserslautern, Germany, 1989. Short version in ECAI'90, p. 688–693.

[Wei91] Christoph Weidenbach. A sorted logic using dynamic sorts. Technical Report MPI-I-91-218, Max-Planck-Institut für Informatik, Im Stadtwald, Saarbrücken, Germany, 1991. Short version in IJCAI'93, p. 60–65.

[Wei94] Christoph Weidenbach. First-order tableaux with sorts. In Krysia Broda and Marcello D'Agostino et al., editors, *TABLEAUX-'94, 3rd Workshop on Theorem Proving with Analytic Tableaux and Related Methods*, pages 247–261. Imperial College of Science Technology and Medicine, TR-94/5, April 1994. To appear in the Bulletin of the IGPL.

MUltlog: an Expert System for Multiple-valued Logics

Gernot Salzer*

Technische Universität Wien, Austria

Abstract. This paper presents the MULTLOG system: given the specification of a multiple-valued logic, it constructs a sequent calculus, a natural deduction system and clause formation rules for the logic. Moreover, we describe optimization techniques based on multiple-valued resolution, which yield a minimal sequent calculus.

1 Introduction

Within the last years multiple-valued logics, introduced in 1920 by the logician Łukasiewicz, have attracted considerable attention by the computer science community due to their potential in the verification of soft- and hardware. This has brought about the necessity for automatizing deduction in these logics. Traditionally, two of the most important approaches to automated deduction are sequent calculi [Gen34] and resolution-based calculi [CL73]. Recently it has been shown (a) that there is a completely automatic procedure for generating sequent calculi for a finitely-valued logic, given the truth tables for its propositional connectives and for its quantifiers, and (b) that such sequent calculi immediately yield natural deduction systems and calculi for the transformation of formulas to clause form for many-valued resolution [BF92, BFZ93].

This paper describes the implementation of the latter results in the MULTLOG system.[1] The first two sections introduce basic concepts concerning multiple-valued logics and sequent calculi. Sections 4 and 5 sketch algorithms for optimizing the introduction rules of multiple-valued operators and quantifiers, respectively. The current implementation as well as future enhancements of MULTLOG are described in the last two sections.

The paper is not intended to be self-contained. Basic knowledge of resolution-style theorem proving (see e.g. [CL73]) and of sequent calculi for multiple-valued logics [BFZ93] is required. For a general survey on automated deduction in multiple-valued logics see [Häh93].

* Current address: Technische Universität Wien, Karlsplatz 13/E185-2, A-1040 Wien/Austria; email: `salzer@logic.tuwien.ac.at`

[1] An early prototype of MULTLOG was already presented in [BFOZ92]. Unfortunately, this implementation turned out to be buggy, and never made it to an official release. The current version of MULTLOG has been written from scratch; the optimizations described in this paper are an original contribution of the author.

2 Specifying Multiple-valued Logics

A (multiple-valued) logic is characterized by the truth functions associated with its propositional connectives and quantifiers. More precicely, if V denotes the set of truth values, then a total function $\tilde{\Box}: V^n \mapsto V$ is associated with each n-ary connective $\Box$, and a total function $\tilde{Q}: (\wp(V) - \{\emptyset\}) \mapsto V$ with each quantifier Q.[2]

In the first-order, finitely-valued case the functions $\tilde{\Box}$ and $\tilde{Q}$ are finite and thus can be specified by finite tables. For practical purposes, however, this method is rather unwieldy: the size of the tables is exponential in the arity of the operators and in the number of truth values. Fortunately, many operators are defined as the least upper or greatest lower bound with respect to some (semi-)lattice ordering on the truth values. Similarly, typical quantifiers like the universal or existential one are generalizations of operators and thus need not be explicitly defined.

Consider for example the following simple logic specified in the input syntax of the MULTLOG system:

```
truth_values{f,u,c,t}.
ordering(diamond, "f < {u,c} < t").
operator(and/2, glb(diamond)).
quantifier(forall, induced_by and/2).
```

The second line defines **diamond** to be the name of the lattice ordering $f < u, c < t$ on the truth values $V = \{f, u, c, t\}$. The binary operator **and** is the greatest lower bound w.r.t. **diamond**, and the quantifier **all** is induced by this operator. E.g., we have **and**$(u, c) = f$ and **all**$(\{u, c, t\}) = f$.

3 Sequent Calculi for Multiple-valued Logics

In the classical case [Gen34] a sequent is of the form $\Gamma_f \mid \Gamma_t$ (usually written as $\Gamma_f \rightarrow \Gamma_t$), where Γ_f and Γ_t are sequences of formulas. $\Gamma_f \mid \Gamma_t$ is true in a given interpretation iff some formula in Γ_f evaluates to false or some formula in Γ_t evaluates to true. In the multiple-valued case sequents contain one sequence of formulas per truth value. For our example sequents take the form $\Gamma_f \mid \Gamma_u \mid \Gamma_c \mid \Gamma_t$; they are true in an interpretation if for some truth value v there is a formula A in Γ_v evaluating to v.

Sequent calculi contain one introduction rule per truth value and operator/quantifier. These rules are a mirror image of the truth functions defining the semantics of the operator or quantifier. Consider for instance the introduction rules for the $\wedge$-operator in classical logic.

$$\frac{\Gamma_f, A, B \mid \Gamma_t}{\Gamma_f, A \wedge B \mid \Gamma_t} \wedge{:}f \qquad \frac{\Gamma_f \mid \Gamma_t, A \quad \Gamma_f \mid \Gamma_t, B}{\Gamma_f \mid \Gamma_t, A \wedge B} \wedge{:}t$$

[2] Quantifiers defined this way are called *distribution quantifiers*. The intuitive meaning is that a quantified formula $(Qx)A(x)$ takes the value $\tilde{Q}(W)$ if the instances $A(d)$ take exactly the elements of W as their value. E.g., the universal quantifier in classical logic can be defined as $\tilde{\forall}(\{t\}) = t$ and $\tilde{\forall}(\{f\}) = \tilde{\forall}(\{t, f\}) = f$.

The first rule can be interpreted as saying that $A \wedge B$ *takes the truth value f iff either A or B is false.* Similarly, the second one reads $A \wedge B$ *takes the value t iff both A and B are true.*[3] Thus, in order to construct the introduction rules, we need for each operator $\Box$ (quantifier Q) and each truth value v a conjunction of disjunctions describing precisely the situations in which the truth function $\tilde{\Box}$ ($\tilde{Q}$) takes the value v. In general such conjunctive forms are not unique. Our goal will be to find optimal ones: the number of conjuncts should be minimal, and in the case of several minimal solutions we prefer those with the smallest total number of literals.

4 Optimized Rules for Operators

We illustrate the algorithm by computing an introduction rule for the operator **and** w.r.t. the truth value c (see Section 2 above). The truth table for **and** is given by

and	f	u	c	t
f	f	f	f	f
u	f	u	f	u
c	f	f	c	c
t	f	u	c	t

In the following, we use F^v as an abbreviation for the proposition *'formula F takes truth value v'*. Our goal is to find a conjunctive form for $\mathbf{and}(A, B)^c$.

The first step is to characterize the situations where $\mathbf{and}(A, B)$ does *not* take the value c:

$$\mathbf{and}(A, B)^c = \neg(\mathbf{and}(A, B)^u \vee \mathbf{and}(A, B)^f \vee \mathbf{and}(A, B)^t) \ .$$

Each of the formulas $\mathbf{and}(A, B)^v$ can be immediately replaced by a disjunctive form. E.g., we have $\mathbf{and}(A, B)^u = (A^u \wedge B^u) \vee (A^u \wedge B^t) \vee (A^t \wedge B^u)$. Moving the negation inward (De Morgan's law) we are almost done:

$$\mathbf{and}(A, B)^c = (\neg A^u \vee \neg B^u) \wedge (\neg A^u \vee \neg B^t) \wedge (\neg A^t \wedge \neg B^u) \wedge \cdots \ .$$

Unfortunately, the negation sign preceding the atomic formulas is part of the meta-language and therefore may not appear in the final sequents. However, exploiting the tautology $F^f \vee F^u \vee F^c \vee F^t$, each literal $\neg F^v$ can be replaced by a disjunction of all atomic formulas F^w satisfying $w \neq v$. Thus we obtain

$$\mathbf{and}(A, B)^c = (A^f \vee A^c \vee A^t \vee B^f \vee B^c \vee B^t) \wedge \cdots \ .$$

This conjunctive form could be immediately translated to an introduction rule. The only drop of bitterness is the size ot the rule obtained that way: it consists of 13 premises, each containing 6 formulas. This is where optimization comes into play: *any* conjunctive form logically equivalent to the above one will do the job. The MULTLOG system employs the following two-step optimization procedure.

[3] Remember that a sequent is a *disjunction* of formulas, whereas the premises of a rule form a *conjunction*.

1. Saturate the conjunctive form obtained by the above method under multiple-valued resolution,[4] tautology elimination and subsumption. With an appropriate strategy this step produces very quickly a much smaller conjunctive form, which is the union of all minimal solutions.[5] In many cases the minimal solution is unique, and the conjunctive form obtained in this step is already minimal.
2. Check all subsets of the clause set obtained in step 1 for completeness. This can be done efficiently by analyzing the history of each clause. Any clause being the only descendant of some original clause belongs to every minimal solution.

Returning to our example, we end up with the optimal introduction rule

$$\frac{\Gamma_f \mid \Gamma_u \mid \Gamma_c, A \mid \Gamma_t, A \quad \Gamma_f \mid \Gamma_u \mid \Gamma_c, A, B \mid \Gamma_t \quad \Gamma_f \mid \Gamma_u \mid \Gamma_c, B \mid \Gamma_t, B}{\Gamma_f \mid \Gamma_u \mid \Gamma_c, \mathbf{and}(A, B) \mid \Gamma_t}\ \mathbf{and}{:}c$$

corresponding to the conjunctive form $(A^c \vee A^t) \wedge (A^c \vee B^c) \wedge (B^c \vee B^t)$.

5 Optimized Rules for Quantifiers

We illustrate the algorithm by computing an introduction rule for the quantifier `all` w.r.t. the truth value c. The truth function is given by

$$\begin{array}{l}\texttt{all}(\{t\}) = t, \quad \texttt{all}(W) = f \text{ for } f \in W \text{ or } \{u, c\} \subseteq W, \\ \texttt{all}(\{c\}) = \texttt{all}(\{c, t\}) = c, \quad \text{and} \quad \texttt{all}(\{u\}) = \texttt{all}(\{u, t\}) = u \ .\end{array}$$

In the following, we use v^+ as an abbreviation for the proposition '*v is in the set of truth values*' and v^- for the opposite assertion.

Our first step is to characterize all sets W such that $\texttt{all}(W)$ does *not* take the value c. E.g., this is the case for the argument $\{u, t\}$; this set can be described by the formula $f^- \wedge c^- \wedge u^+ \wedge t^+$. Doing this for all W satisfying $\texttt{all}(W) \neq c$, we obtain a disjunctive normal form in the propositional variables f, u, c and t. Furthermore, we add $f^- \wedge c^- \wedge u^- \wedge t^-$ representing the empty set, which has the status of a *don't-care* value. In many cases this additional formula leads to a smaller introduction rule, which is equivalent to setting $\texttt{all}(\emptyset) \neq c$. If it is still present after the optimization process, it may be dropped (corresponding to $\texttt{all}(\emptyset) = c$).

As a first simplification we apply the Quine-McCluskey procedure—or alternatively, resolution combined with tautology elimination and subsumption—to the disjunctive form obtained above, which yields the formula $u^+ \vee f^+ \vee c^-$. Its negation, $u^- \wedge f^- \wedge c^+$, captures all sets W satisfying $\texttt{all}(W) = c$. Next we

[4] Multiple-valued resolution is analogous to two-valued resolution. Two clauses C, D may be resolved upon any literals $F^v \in C$ and $F^w \in D$ satisfying $v \neq w$. For details see [BF92].

[5] In fact, for the two-valued case one particular strategy coincides with the well-known Quine-McCluskey algorithm.

replace every literal v^+ by $(\exists x)A^v(x)$ and every literal v^- by $(\forall x)\neg A^v(x)$. By Skolemizing the resulting first order formula we get rid of the quantifiers. The negation signs are eliminated in the same way as in the last section. We obtain

$$(A^f(x)\lor A^c(x)\lor A^t(x)) \land (A^u(x)\lor A^c(x)\lor A^t(x)) \land A^c(a) \ .$$

The last two steps are those of Section 4: saturation under resolution etc., and selection of minimal subsets. The only difference is that in addition to resolution, subsumption and tautology elimination we need factorization or condensation. The above clause set collapses to $(A^c(x)\lor A^t(x))\land A^c(a)$, which translates to the introduction rule

$$\frac{\Gamma_f \mid \Gamma_u \mid \Gamma_c, A(\alpha) \mid \Gamma_t, A(\alpha) \quad \Gamma_f \mid \Gamma_u \mid \Gamma_c, A(\tau) \mid \Gamma_t}{\Gamma_f \mid \Gamma_u \mid \Gamma_c, (\texttt{all}\, x)A(x) \mid \Gamma_t}\ \texttt{all}{:}c$$

where α is an eigenvariable and τ is a term variable.

6 The MULTLOG system

The MULTLOG system automatizes the process of constructing minimal sequent calculi. Its input consists of an ordinary text file containing the specification of a multiple-valued logic; an example for such a specification has been given in Section 2. To make the input more user-friendly there are three different types of editors, each of them producing such a text file.[6] The kernel is written in standard Prolog and has been tested with SICStus Prolog and BinProlog on Unix workstations as well as on PCs. The output of MULTLOG is in the form of a scientific paper (written in LaTeX): it contains an optimized sequent calculus, a natural deduction system and clause formation rules for that logic. MULTLOG is available via anonymous ftp from host `logic.tuwien.ac.at`, directory `pub/multlog`.

7 Future Developements

Future versions of MULTLOG will include the following enhancements:

- MULTLOG will be able to construct cut elimination algorithms for multiple-valued logics [BFZ94] and to include corresponding cut elimination theorems into its output.
- MULTLOG will include knowledge about already known logics. Each newly input logic will be compared against this database and appropriate comments and references will be added to the paper.
- MULTLOG will be linked to an automatic theorem prover, such that the clause formation rules for a particular logic are not just included into the paper, but also result in a clausal theorem prover for that logic.

[6] The first one (by A. Leitgeb) is written in Tcl/Tk and runs under Unix and X-Windows. The second one (by W. Nix) is written in C for PCs under DOS. The last one (by M. Schranz) is written in HTML/Perl, providing a WWW-interface; this makes it possible to use MULTLOG without installing it on ones own machine.

References

[BF92] M. Baaz and C. G. Fermüller. Resolution for many-valued logics. In Voronkov [Vor92], pages 107–118.

[BFOZ92] M. Baaz, C. G. Fermüller, A. Ovrutcki, and R. Zach. MULTLOG: A system for axiomatizing many-valued logics. In Voronkov [Vor92], pages 345–347.

[BFZ93] M. Baaz, C. G. Fermüller, and R. Zach. Systematic construction of natural deduction systems for many-valued logics. In *Proc. 23rd International Symposium on Multiple-valued Logic*, pages 208–213. IEEE Press, 1993.

[BFZ94] M. Baaz, C. G. Fermüller, and R. Zach. Elimination of cuts in first-order finite-valued logics. *J. Inform. Process. Cybernet. EIK*, 29(6):333–355, 1994.

[CL73] C. L. Chang and R. C. T. Lee. *Symbolic Logic and Mechanical Theorem Proving.* Academic Press, 1973.

[Gen34] G. Gentzen. Untersuchungen über das logische Schließen, I–II. *Math. Z.*, 39:176–210, 405–431, 1934.

[Häh93] R. Hähnle. *Automated Deduction in Multiple-valued Logics.* Clarendon Press, Oxford, 1993.

[Vor92] A. Voronkov, editor. *Logic Programming and Automated Reasoning (LPAR'92)*, LNCS 624 (LNAI). Springer-Verlag, 1992.

A Fundamental Problem of Mathematical Logic

Jan Krajíček

Mathematical Institute of the Academy of Sciences
Žitná 25, Praha 1, 115 67, Czech Republic
krajicek@earn.cvut.cz

A fundamental open problem of mathematical logic and simultaneously the main problem of computational complexity theory is the following one.

> **Problem (an informal formulation):** *What is the most efficient way of deciding whether a propositional formula is satisfiable or not?*

It is a problem addressing a quite rudimentary topic at a fundamental level and having far-reaching connections to other parts of logic, mathematics and philosophy. Moreover, the research stimulated by this problem consistently generates interesting results and intrigued notions.

The $\mathcal{P} =_? \mathcal{NP}$ is the most famous technical version of the problem, see [15]. It asks whether there can be an algorithm deciding the satisfiability of formulas and running in time polynomial in the size of the formula. The size of a formula (or a proof) is the total number of symbols in it. This is proportional to the number of bits needed to encode the formula for a machine.

The main question about propositional calculus is whether there is a propositional proof system in which every tautology has a short proof (of size polynomial in the size of the tautology).

The main problem in bounded arithmetic is whether it is a finitely axiomatizable theory. This appears to be rather remote from the problem considered above but we shall see that it is not.

In the following five sections we briefly explain in a not too technical language the problems in all three areas (Sections 1 - 3) and the connections between them (Sections 4 and 5). The last section describes some recent work and formulates a technical problem which is currently investigated.

The paper is aimed at non-specialists. The text is accompanied by extensive references to original sources. However, most of the material can be found in the monograph [26] which an interested reader may use as an entrance into the field.

1 Computational Complexity

SAT is the set of satisfiable propositional formulas in the De Morgan language 1 (truth), 0 (false), $\vee$ (or), $\wedge$ (and) and $\neg$ (not). The set *SAT* is in $\mathcal{NP}$, the

This is an outline of my *Collegium Logicum* lecture *Bounded arithmetic, propositional logic and complexity theory* delivered in the Kurt Godel Society on December 12, 1994.

class of sets accepted by a non-deterministic polynomial-time Turing machine; the machine simply guesses a truth assignment and accepts only if it is indeed a satisfying assignment.

$TAUT$ is the set of tautologies in the same language. Clearly, $\phi \in SAT$ iff $\neg\phi \notin TAUT$. Hence the complement $\backslash TAUT$ of $TAUT$ is in $\mathcal{NP}$. That is, $TAUT$ itself is in $co\mathcal{NP}$, the class of the complements of $\mathcal{NP}$-sets. Cook's theorem [15] says that

1. $\mathcal{NP} = co\mathcal{NP}$ iff $SAT \in co\mathcal{NP}$ iff $TAUT \in \mathcal{NP}$.
2. $\mathcal{NP} = \mathcal{P}$ iff $SAT \in \mathcal{P}$ iff $TAUT \in \mathcal{P}$.

This is the so called $\mathcal{NP}$-completeness of SAT and the $co\mathcal{NP}$-completeness of $TAUT$.

The trivial algorithm checking all possible truth assignments to the atoms in the formula is, essentially, the most efficient algorithm known (in other words: nothing is known). The number of truth assignments is 2^n where n is the number of atoms. As n might be close to the size of the formula this means that the trivial algorithm runs almost in exponential time.

Considerably more is known about the mutual relation between SAT and other decision problems in combinatorics or number theory; these are the $\mathcal{NP}$-completeness results, see [19]. There are also proofs that algorithms from some restricted classes of algorithms are not more efficient (in a substantial way) than the trivial algorithm. Here belong lower bounds to the circuit-size for various circuit models: constant depth ([1, 18, 55, 46, 54]) and monotone ([45, 6, 5]); see [9] (or [26, Chpt.3]) for an overview.

2 Propositional Logic

F (for a *Frege system*) is the often used propositional calculus in the language of SAT based on a finite number of axiom schemes and the modus ponens:

$$\frac{\phi \quad \neg\phi \vee \psi}{\psi}$$

as the only inference rule. SF (for F with *substitution*) is F augmented by the substitution rule:

$$\frac{\phi(p_1, \ldots, p_n)}{\phi(\psi_1, \ldots, \psi_n)}$$

allowing to infer in one step any substitution instance of an already derived formula (see [26, Chpt.4]).

It is a fundamental problem whether every tautology ϕ admits a proof in F or in SF of size polynomial in the size of ϕ. No non-trivial lower bounds are known (but non-trivial upper bounds for particular ϕ are known, e.g. [11] or [26, Chpt.13]).

Assume that every tautology would indeed admit polynomial size F-proofs. Then $TAUT \in \mathcal{NP}$ and hence $\mathcal{NP} = co\mathcal{NP}$. This is because a machine which guesses a polynomial size string and then accepts iff the string is an F-proof of the formula would be a non-deterministic acceptor of $TAUT$ running in polynomial time. With a suitable definition of a propositional proof system (as in [17]) the statement that $\mathcal{NP} = co\mathcal{NP}$ is actually equivalent to the statement that there is a propositional proof system in which all tautologies have polynomial size proofs (see [26, Chpt.4]. In fact, if such a proof system exists it could be formulated as an extension of SF by a polynomial-time set of tautologies as new axioms, see [28] or [26, Chp.14].

Obviously then, one expects that no proof system admits polynomial size proofs for all tautologies (as $\mathcal{NP} \neq co\mathcal{NP}$ is a generally accepted conjecture). This was, however, demonstrated only for some particular subsystems of F. Most notable examples are resolution ([20]), constant-depth systems ([2, 24, 33, 44]) and constant-depth systems augmented by additional axiom-schemes expressing various combinatorial principles ([3, 50, 8, 4, 7, 51]), see [26, Chpts.4 and 12]. Nothing is known for the constant-depth systems if the language contains also the equivalence connective $\equiv$. To prove a lower bound for such systems is the most accessible problem at present.

All known lower bounds fit into one methodological framework, the so called *partial boolean valuations* defined in [25]. It was proved there that, in principle, a valid lower bound for SF can be proved by a construction of a suitable partial boolean algebra and a map of formulas into it. However, such a construction is quite complicated even for the restricted cases considered above, see [26, Chpt.13].

3 Bounded Arithmetic

Bounded arithmetic is a subsystem of Peano Arithmetic in which the induction axioms are restricted to bounded formulas only. It was proposed by [38]. Several other systems, differing in the language and in the particular class of bounded formulas for which the induction is assumed, were defined and studied since then ([16, 39, 40, 41, 10]). Here we shall consider only the theory S^1_2 of [10], perhaps the most important one in the present context (see [26, Chpt.5]).

The language of S^1_2 is a bit richer then the language of Peano Arithmetic: $0, 1, +, \cdot, =, \leq$. In addition, it has three extra function symbols $|x|$ (the length of the dyadic numeral of x, $|x| = \lceil \log_2(x+1) \rceil$ for $x > 0$), $\lfloor \frac{x}{2} \rfloor$ and $x \# y$ (standing for $2^{|x| \cdot |y|}$). This language was motivated by the initial functions used for the machine-independent characterization of the polynomial-time functions by [14].

The class of bounded formulas in this language is stratified into levels $\Sigma^b_1 \subseteq \Sigma^b_2 \subseteq \ldots$ in a manner analogous to the arithmetical hierarchy $\Sigma^0_1 \subseteq \Sigma^0_2 \subseteq \ldots$ The first and most straightforward connection of bounded arithmetic to complexity theory is that bounded formulas define exactly those subsets of $\mathbf{N}$ which are in the polynomial-time hierarchy $\mathcal{PH}$ of [37], and the Σ^b_i-formulas define the sets from the i th-level of $\mathcal{PH}$. In particular, the Σ^b_1-formulas define exactly the $\mathcal{NP}$-sets. See also [26, Chpt.3].

S^1_2 is axiomatized by a finite number of axioms codifying the basic recurrence properties of the functions in the language (the set $BASIC$ in [10]), and the following form of the induction axiom:

$$A(0) \wedge \forall x(A(x) \rightarrow A(x+1)) \longrightarrow \forall x A(|x|)$$

accepted for all Σ^b_1-formulas A. This form of induction is called the *length induction.*

The main theorem of [10], the so called *witnessing theorem*, stretches the correspondence between bounded formulas and sets from $\mathcal{PH}$ further. Namely, a function is Σ^b_1-definable in S^1_2 iff it is a polynomial-time function (and analogously for $i > 1$). See also [26, Chpt.7].

A particular instance of the finite axiomatizability problem (see [26, Chpt.10]) is whether already S^1_2 proves the length-induction for all bounded formulas and not just for Σ^b_1-formulas. Obviously, if for every bounded formula $A(x)$ there is a Σ^b_1-formula $B(x)$ such that S^1_2 proves $\forall x(A(x) \equiv B(x))$, then the answer is yes. This hypothesis is abbreviated by the phrase that S^1_2 proves that $\mathcal{PH} = \mathcal{NP}$. More interestingly, a form of the converse also holds: if S^1_2 entails the length-induction for all bounded formulas then the polynomial-time hierarchy collapses (to $\mathcal{PH} = \Sigma^p_2$ by [32]) and collapses even provably in S^1_2 (see [12, 56]).

The finite axiomatizability problem thus asks whether there is a model of bounded arithmetic in which $\mathcal{PH}$ does not collapse (and, in particular, in which there is no proof system in which all tautologies would have polynomial-size proofs in any proof system).

4 Propositional Logic and Bounded Arithmetic

Let T be a theory capable of coding finite strings. For example, $T = S^1_2, PA$ or ZFC. For simplicity assume that the language of T contains the language of S^1_2 and that $T \supseteq S^1_2$. There is a Π^b_1-formula (a negation of a Σ^b_1-formula) $Taut(x)$ such that for every propositional formula ϕ, $Taut(\lceil\phi\rceil)$ holds about the string $\lceil\phi\rceil$ encoding ϕ iff ϕ is a tautology. $Taut(\lceil\phi\rceil)$ is a bounded sentence and hence, if true, can be proved in T. This T-proof can be interpreted as a proof of ϕ in a propositional proof system P_T determined by T. Hence a theory can be thought of as a propositional proof system.

Let $A(x)$ be an arithmetical formula and n a number. Any string $\epsilon = (\epsilon_0, \ldots, \epsilon_{n-1}) \in \{0,1\}^n$ defines a number $\tilde{\epsilon} := \sum_{i<n} \epsilon_i \cdot 2^i$ and so $A(x)$ defines a Boolean function $f_{A,n}$:

$$f_{A,n}(\epsilon) := \begin{cases} 1 \text{ if } A(\tilde{\epsilon}) \text{ holds} \\ 0 \text{ otherwise} \end{cases}$$

Moreover, if $A(x)$ is a Π_1^b-formula then for every n there is a propositional formula $\|A\|^n$ in atoms $p_0, \ldots, p_{n-1}, q_1, \ldots, q_t$ of size polynomial in n such that:

$$f_{A,n}(\epsilon) = 1 \text{ iff } \|A\|^n(p_i/\epsilon_i, \overline{q}) \in TAUT .$$

This is a corollary of the $co\mathcal{NP}$-completeness of $TAUT$, see [16, 28] or [26, Chpt.9].

Let P be a proof system. With this translation a P-proof of the formula $\|A\|^n$ can be interpreted as a proof of the arithmetic sentence $\forall x, |x| \leq n \rightarrow A(x)$ in a theory T_P determined by P. Hence a propositional proof system can be considered as an arithmetical theory.

We have the following correspondence between arithmetic and propositional proof systems:

1. If T proves $\forall x A(x)$, where A is a Π_1^b-formula, then the formulas $\phi_n := \|A\|^n$ are tautologies and there is a polynomial-time function constructing from $1 \ldots 1$ (n-times) a P_T-proof of ϕ_n.
2. If ϕ_n is a sequence of tautologies having polynomial-size P-proofs definable in T_P, then T_P proves $\forall x Taut(\ulcorner \phi_x \urcorner)$. If $\phi_n = \|A\|^n$ then T_P proves also $\forall x A(x)$.

Moreover (see [28] or [26, Chpt.9]):

3. T and $T_{(P_T)}$ prove the same $\forall \Pi_1^b$-formulas, and P and $P_{(T_P)}$ have proofs of same sizes (up to a polynomial increase).

What is most striking, however, is that there are pairs (T, P) of a *natural* theory T and a *natural* propositional proof system P enjoying such a correspondence. *Natural* means that the systems were defined and studied prior to such a correspondence.

The best example is the pair (S_2^1, SF). In particular, any finite combinatorial principle (like the pigeonhole principle and similar), formalizable by a $\forall \Pi_1^b$-formula, is provable in S_2^1 iff the tautologies ϕ_n, encoding that the principle holds in structures of size n, have uniformly definable polynomial-size SF-proofs.

The following papers are related [16, 40, 28, 29, 34, 35, 25], or see [26, Chpt.9].

5 Further Connections to the Complexity Theory

The connections explained in Sections 3 and 4 establish a strong link of bounded arithmetic to complexity theory. The deepest among those is, in my opinion, the relation between the finite axiomatizability problem and the collapse of the polynomial-time hierarchy from [32]. This problem can be viewed also from the following angle.

It often happens that an important problem in a field of mathematics inspires two types of research. The research of the first type attempts to answer weaker variants of the original problem. The research of the second type tries to verify the conjectured solution of the problem in a class of structures similar to the structure intended in the original formulation of the problem. Both seem to be quite essential for a development of methods needed for a full solution of the problem.

Algebraic geometry and number theory provide good examples. Hilbert's 10th problem for the field of rationals (about algorithmic solvability of Diophantine equations over $\mathbf{Q}$) inspires decision procedures for classes of equations and unsolvability results for bigger classes of first-order formulas over $\mathbf{Q}$, but also proofs of Matijasevic theorem [36] for various function fields over a finite field which bear relation to $\mathbf{Q}$, see [42, 43]. Another example is provided by the famous Riemann hypothesis. Besides partial results in its directions, analogous statements were formulated and proved for (various curves over) a finite field, see [52].

The intended structure for the conjecture that $\mathcal{P} \neq \mathcal{NP}$ (or that $\mathcal{PH}$ does not collapse in general) is the set of natural numbers $\mathbf{N}$ and the structures where the involved notions make a very good sense are models of bounded arithmetic. Thus to show that $\mathcal{P} \neq \mathcal{NP}$ in a class of models of bounded arithmetic appears to be a major research task, complementing the program of Boolean complexity to prove lower bounds to circuit-size for various restricted circuit models. The problem to establish the conjecture that $\mathcal{PH}$ does not collapse for a class of models of S_2 is just the finite axiomatizability problem.

Further connections to Boolean complexity are provided by [47] who showed that methods used at present for circuit-size lower bounds can be formalized in a suitable systems of bounded arithmetic, see also [26, Chpt.15].

6 A Current Research

The following two problems, two sides of the same coin by section 4, were currently investigated ([24, 48, 49, 27, 31]) and are not yet fully settled.

1. Let P be a propositional proof system. Estimate the circuit-size of an interpolant of an implication from the minimum size of its P-proof.
2. Let T be a theory and assume that T proves that two $\mathcal{NP}$-sets (represented by Σ_1^b-formulas) U and V are disjoint. Estimate the complexity (circuit-size) of a set separating the two sets.

If T proves $U(x) \to \neg V(x)$, where $U(x)$ and $V(x)$ are two Σ_1^b-formulas, then the implications $\neg\|\neg U\|^n \to \|\neg V\|^n$ have polynomial size P_T-proofs, and the interpolants I_n for them define the set:

$$W = \bigcup_n \{\epsilon \in \{0,1\}^n \mid I_n(\epsilon)\}$$

separating $U = \{x \mid U(x)\}$ and $V = \{x \mid V(x)\}$: $U \subseteq W \wedge W \cap V = \emptyset$. Hence an upper bound for the first problem for $P = P_T$ yields an upper bound for the second one. In the opposite direction a similar reduction works. Take $T \supseteq T_P$ in which the P-proofs of the implications $\phi_n(\overline{p},\overline{q}) \rightarrow \psi_n(\overline{p},\overline{r})$ are definable. Then T proves the disjointness of the sets $U = \bigcup_n \{\epsilon \in \{0,1\} \mid \exists\delta\phi_n(\epsilon,\delta)\}$ and $V = \bigcup_n \{\epsilon \in \{0,1\} \mid \exists\xi\neg\psi_n(\epsilon,\xi)\}$, and every set separating U from V yields interpolants for the implications.

The problem is relevant to the lower bound problem for propositional logic as follows ([24]). Assume that two $\mathcal{NP}$-sets cannot be separated by a set of small circuit-size complexity. Assume also that a proof system P admits a polynomial upper bound in the first problem. Then we have a lower bound for P. Namely, the implications determined by the two sets cannot have short P-proofs as otherwise the two sets could be separated by a low complexity set (by the assumption about the existence of a polynomial upper bound for problem 1) contradicting the assumption about the inseparability of U and V.

This problem is, to a large extent, still open. The reader may consult [27] for up-to-day (January 1995) information.

Acknowledgement: I thank V. Švejdar for comments on the manuscript of this paper.

References

1. Ajtai, M. (1983) Σ_1^1 - formulae on finite structures, *Annals of Pure and Applied Logic*, **24** : 1-48.
2. —— (1988) The complexity of the pigeonhole principle, in: *Proc. IEEE 29th Annual Symp. on Foundation of Computer Science*, pp. 346-355.
3. —— (1990) Parity and the pigeonhole principle, in: *Feasible Mathematics*, eds. S.R. Buss and P.J. Scott, pp.1-24. Birkhauser.
4. —— (1994) The independence of the modulo p counting principles, in: *Proceedings of the 26th Annual ACM Symposium on Theory of Computing*, pp.402-411. ACM Press.
5. Alon, N., and Boppana, R. (1987) The monotone circuit complexity of boolean functions, *Combinatorica*, **7(1)** : 1-22.
6. Andreev, A. E. (1985) On a method for obtaining lower bounds for the complexity of individual monotone functions (in Russian), *Doklady AN SSSR*, **282(5)** : 1033-1037.
7. Beame, P., Impagliazzo, R., Krajíček, J., Pitassi, T., and Pudlák, P. (1994) Lower bounds on Hilbert's Nullstellensatz and propositional proofs, submitted.
8. Beame, P., and Pitassi, T. (1993) Exponential separation between the matching principles and the pigeonhole principle, preprint.
9. Boppana, R., and Sipser, M. (1990) Complexity of finite functions. in:*Handbook of Theoretical Computer Science*, ed. J. van Leeuwen, pp.758-804.
10. Buss, S. R. (1986) Bounded Arithmetic. Naples, Bibliopolis. (Revision of 1985 Princeton University Ph.D. thesis.)
11. —— (1987) The propositional pigeonhole principle has polynomial size Frege proofs, *J. Symbolic Logic*, **52**: 916-927.
12. —— (1993) Relating the bounded arithmetic and polynomial time hierarchies, manuscript.

13. Clote, P., and Krajíček, J. (1993) Open problems, in: *Arithmetic, Proof Theory and Computational Complexity*, eds.P. Clote and J. Krajíček, pp.1-19, Oxford Press.
14. Cobham, A. (1965) The intrinsic computational difficulty of functions, in : *Proc. Logic, Methodology and Philosophy of Science*, ed. Y. Bar-Hillel, p. 24-30, North-Holland.
15. Cook, S. A. (1971) The complexity of theorem proving procedures, in: *Proc. 3rd Annual ACM Symp. on Theory of Computing*, pp. 151-158. ACM Press.
16. —— (1975) Feasibly constructive proofs and the propositional calculus, in: *Proc. 7th Annual ACM Symp. on Theory of Computing*, pp. 83-97. ACM Press.
17. Cook, S. A., and Reckhow, A. R. (1979) The relative efficiency of propositional proof systems, *J. Symbolic Logic*, **44(1)**:36-50.
18. Furst, M., Saxe, J. B., and Sipser, M. (1984) Parity, circuits and the polynomial-time hierarchy, *Math. Systems Theory*, **17**: 13-27.
19. Garey, M.R., and Johnson, D. S. (1979) Computers and intractability. New York, W.H. Freeman and Co..
20. Haken,A. (1985) The intractability of resolution, *Theoretical Computer Science*, **39**:297-308.
21. Hastad, J. (1989) Almost optimal lower bounds for small depth circuits. in: *Randomness and Computation*, ed. S. Micali, Ser. Adv. Comp. Res. **5**: 143-170. JAI Press.
22. Krajíček, J. (1989) Speed-up for propositional Frege systems via generalizations of proofs, *Commentationes Mathematicae Universitas Carolinae*, **30(1)**:137-140.
23. —— (1993) Fragments of bounded arithmetic and bounded query classes, *Transactions of the A.M.S.*, **338(2)** : 587-598.
24. —— (1994) Lower bounds to the size of constant-depth propositional proofs, *Journal of Symbolic Logic*, **59(1)**, pp.73-86.
25. —— (1995) On Frege and Extended Frege proof systems, in: *Feasible Mathematics II*, eds. P. Clote and J. Remmel, Birkhauser, pp.284-319.
26. —— (1994) *Bounded arithmetic, propositional logic and complexity theory*, Cambridge University Press, in print.
27. —— (1995) Interpolation theorems, lower bounds for proof systems and independence results for bounded arithmetic, preprint.
28. Krajíček, J., and Pudlák, P. (1989) Propositional proof systems, the consistency of first order theories and the complexity of computations, *J. Symbolic Logic*, **54(3)**:1063-1079
29. —— (1990) Quantified propositional calculi and fragments of bounded arithmetic, *Zeitschrift f. Mathematikal Logik u. Grundlagen d. Mathematik*, **36**:29-46.
30. —— (1990) Propositional provability in models of weak arithmetic, in: *Computer Science Logic (Kaiserlautern, Oct. '89)*, eds. E. Boerger, H. Kleine-Bunning and M.M. Richter, LNCS 440, pp.193-210. Springer-Verlag.
31. —— (1995) Some consequences of cryptographical conjectures for S_2^1 and EF, submitted.
32. Krajíček, J, Pudlák, P, and Takeuti, G. (1991) Bounded arithmetic and the polynomial hierarchy, *Annals of Pure and Applied Logic*, **52**: 143-153.
33. Krajíček, J.,Pudlák, P. and Woods, A. (1991) Exponential lower bound to the size of bounded depth Frege proofs of the pigeonhole principle, submitted.
34. Krajíček, J., and Takeuti, G. (1990) On bounded $\sum_1^1$-polynomial induction, in: *Feasible Mathematics*, eds. S.R. Buss and P.J. Scott, pp.259-280. Birkhauser.
35. —— (1992) On induction-free provability, *Annals of Mathematics and Artificial Intelligence*, **6**: 107-126.

36. Matijasevic, Y. (1970) Enumerable sets are Diophantine, *Doklady AN SSSR*, **191**, pp.279-282.
37. Meyer, A., and Stockmeyer, L. (1973) The equivalence problem for regular expressions with squaring requires exponential time, in: *Proc. IEEE 13th Symp. on Switching and Automata Theory*, pp. 125-129.
38. Parikh, R. (1971) Existence and feasibility in arithmetic, *Journal of Symbolic Logic*, **36**, pp.494-508.
39. Paris, J., and Wilkie, A. J. (1981) Δ_0 sets and induction, in: *Proc. of the Jadwisin Logic Conf.*, Poland, pp.237-248. Leeds University Press.
40. —— (1985) Counting problems in bounded arithmetic, in: *Methods in Mathematical Logic*, LNM 1130, pp.317-340. Springer.
41. —— (1987) On the scheme of induction for bounded arithmetic formulas, *Annals of Pure and Applied Logic*, **35**, pp.261-302.
42. Pheidas, T. (1991) Hilbert's tenth problem for fields of rational functions over finite fields, *Inventiones Mathematicae*, **103**, pp.1-8.
43. —— (1994) Extensions of Hilbert's tenth problem, *Journal of Symbolic Logic*, **59(2)**, pp.372-397.
44. Pitassi, T., Beame, P., and Impagliazzo, R. (1993) Exponential lower bounds for the pigeonhole principle, *Computational Complexity*, **3**, pp.97-308.
45. Razborov, A. A. (1985) Lower bounds on the monotone complexity of some Boolean functions, *Soviet Mathem. Doklady*, **31**, pp.354-357.
46. —— (1987) Lower bounds on the size of bounded depth networks over a complete basis with logical addition, *Matem. Zametki*, **41(4)** : 598-607.
47. —— (1994) Bounded arithmetic and lower bounds in Boolean complexity, in : *Feasible Mathematics II*, eds. P. Clote and J. Remmel, Birkhauser, pp.344-386.
48. —— (1994) Unprovability of lower bounds on the circuit size in certain fragments of bounded arithmetic, *Izvestiya of the R.A.N.*, to appear.
49. —— (1994) On provably disjoint NP-pairs, preprint.
50. Riis, S. (1993) Independence in bounded arithmetic, PhD. Thesis, Oxford University.
51. —— (1994) *Count(q)* does not imply *Count(p)*, preprint.
52. Shafarevich, I. R. (1974) Basic algebraic geometry. Springer.
53. Smale, S. (1992) Theory of computation, in: *Mathematical Research Today and Tomorrow*, eds. C. Casacuberta and M. Castellat, pp.59-69. Springer.
54. Smolensky, R (1987) Algebraic methods in the theory of lower bounds for Boolean circuit complexity, in: *Proc. 19th Ann. ACM Symp. on Theory of Computing*, pp. 77-82.
55. Yao, Y. (1985) Separating the polynomial-time hierarchy by oracles, in: *Proc. 26th Ann. IEEE Symp. on Found. of Comp. Sci.*, pp. 1-10.
56. Zambella, D. (1994) Notes on polynomially bounded arithmetic, preprint.

On the lengths of proofs of consistency

a survey of results

Pavel Pudlák

Mathematical Institute
Academy of Sciences
Prague, Czech Republic

Preface. This article is essentially a part of my thesis for the degree DrSc (Doctor of Sciences). Therefore it mainly surveys my articles [42, 43, 44, 29, 30, 45, 23], and it is structured according to the requirements for such theses. I made only minor changes in the original text and added a few further references. Since Gödel's main achievement concerns the problem of consistency and some of the problems that I am going to describe had been considered by him, I think that it is appropriate to publish this article in Gödel Society.

1 Historical remarks

The question that we are going to consider is interesting *per se* and is related to some more practical questions, especially in complexity theory, but the original motivation for it comes from foundational studies. Among the variety of streams in foundations of mathematics, the one which had the biggest influence and which very much determined later development of mathematical logic was Hilbert's *formalism.* It was an attempt to reduce mathematical theories to *logical theories*, which in turn are defined using a formal logical calculus. In this way the obscure concepts of *intuition, meaning, etc.* are eliminated. But then the consistency of a theory, which is a condition *sine qua non*, loses any support, and hence *must be proved.*

Apparently Hilbert hoped that eventually it will be possible to prove consistencies of all mathematical theories (as they can be formalized in set theory, it would suffice to do it for set theory). This program received a serious blow, before there was any significant progress in proving such consistencies.[1] The result, which in fact essentially destroyed it, is what is nowadays referred to as *Gödel's Second Incompleteness Theorem*, which says that the consistency of a theory T cannot be proved in T itself. We shall denote the sentence saying that T is consistent by Con_T. If we restrict the class of theories to some natural systems, say theories which are subsets of the true sentences in the standards model of arithmetic $\mathbf{N}$, then the relation $T \vdash Con_S$ (T proves Con_S) is a strict partial ordering. Hilbert's idea was to use some *finitistic* arguments for the proofs of consistency, i.e. to use simple combinatorial arguments. However, at the same time he assumed that any mathematical argument can be a part of some theory.

[1] The famous proof of consistency of the first order arithmetic of Gentzen [17] was published several years after Gödel's [19].

This implies that we can prove Con_T only in some "stronger" theory; thus if we doubt in the consistency of T, we can only reduce it to the consistency even more doubtful. Of course, for some T, we can reduce the consistency of T, which uses seemingly strong statements, to a theory for which we have more intuitive reasons to believe to be consistent. But then we are back at the beginning – we refer to intuition.

A possible way out is to give up absolute consistency and consider only practical consistency, i.e. the question whether we can ever encounter a contradiction, if we use only small proofs. One can argue that the size of the proofs that men will ever consider is bounded or that the universe is finite etc. At first glance it seems that the restriction to proofs of a size bounded by some constant n makes a big difference. In any case, it is clear that now we do not need a strong theory to prove the consistency up to n, since the procedure of checking all such proofs can be described in a weak theory. The problem is, however, that this may be exponentially longer than n; put otherwise the question is now, to find a *feasible* proof that all feasible proofs are consistent. Kreisel mentioned a discussion with Gödel, where Gödel talked about this problem.

The only attempt of actually proving such a consistency result that I know of, is a paper by Esenin-Volpin [12]. He claimed that he proved the consistency of a set theory up to some large number n using a finitistic argument. Unfortunately his arguments were obscure and I have not met anybody who could explain to me, what actually Esenin-Volpin proved.

These questions influenced several directions of research. We shall mention some results related to the consistency problem.

The consistency problem was an initiating idea for studying inconsistent theories. The basic result about "useful" inconsistent theories is due to Parikh [38], who constructed theories which are inconsistent, but where any proof of contradiction is extremely long, and showed some nice properties of such theories. For further results see [11, 16, 35].

Another problem connected with the result of Gödel was the question, for how weak theories the second incompleteness theorem is true. Bezboruah and Shepherdson [5] proved the second incompleteness theorem in $\mathbf{Q}$, which is one of the weakest arithmetical theories. The result was in certain sense problematic: if the theory is so weak, does the particular formalization of $Con_{\mathbf{Q}}$ really mean what was intended? A solution, which will be presented below, is to define an initial segment J of the numbers in $\mathbf{Q}$, which is an inner model of a stronger theory T and prove that it is consistent to assume that a proof of contradiction from axioms of $\mathbf{Q}$ is encoded by a number which is already in J. Since T is strong, the meaning of $Con_{\mathbf{Q}}$ is not so ambiguous.

Let us observe that this naturally leads to questions about the *lengths* of proofs of contradictions and this in turn can be investigated by computing the lengths of proofs of statements about finite consistency, as we shall see below.

The finite consistency of T is the statement, that *there are no proofs of contradiction of length n in T*, where n is some positive integer. We shall denote this sentence by $Con_T(n)$. The *length of a proof* will always mean the length of the sequence which represents the proof. Clearly, if n is a concrete (standard) integer and T is reasonably strong (in fact already $\mathbf{Q}$ suffices), then $T \vdash Con_T(n)$. The question *what is the minimal length of a proof of* $Con_T(n)$ has been considered for the first time by Harvey Friedman [14]. He proved a lower bound of the form n^ε, for a constant $\varepsilon > 0$. His paper was almost unknown, since it existed only in a preprint form so I rediscovered this result several years later [43].[2]

Around 1980 a new field in logic became popular – weak fragments of arithmetic, nowadays called *Bounded Arithmetic.*[3] The first such system was $I\Delta_0$ which is the usual first order arithmetic (Peano Arithmetic) where induction is postulated only for bounded formulas; later a whole variety of systems has been introduced. This line of research was pursued at the beginning mainly by the Manchester school, Paris, Wilkie and others, later it was very influenced by Buss's book [6]. Though it has a different motivation, this topic is related to the consistency proofs too. Firstly, new more efficient ways of formalizing syntax have been developed. The bounds on the lengths of proofs of finite consistencies $Con_T(n)$ are very much based on such techniques. Secondly, Paris and Wilkie realized that restricted forms of consistency are useful for separations of weak fragments of arithmetic [56]. (They will be defined below.)

In spite of the fact that the importance of the problems about the lengths of proofs has been realized very early, see e.g. Gödel's [20], the literature on this field was not very rich until recently. Kreisel was probably the main proponent of this field. His student Statman [20] proved that the size of proofs after cut-elimination cannot be bounded by an elementary function of the original size. His famous conjecture about generalizations of proofs in Peano Arithmetic has been proved only in some special cases (see e.g. [3]). Another line of research has been started by Cook [9] who was the first to realize the importance of the lower bounds on the lengths of proofs in propositional calculus and a relation of this problem to equational and first order theories. In this field there was quite a significant progress recently; to name just a few [10, 31, 1, 4, 2, 28].

After the famous result of Paris and Harrington [41], who proved independence of a concrete mathematical statement from Peano Arithmetic, we started to study fragments of arithmetic with an intention to prove eventually independence of some simpler (from the point of view of quantifier complexity)[4] sentences, in particular we hoped to prove independence of some problems in complexity theory. After more than 15 years of experience it has become clear that it will be necessary to develop the theory much further in order to obtain such results. Presently we are able to show only so called *oracle results*, which means that we can show independence of a certain combinatorial principle for

[2] I have presented it on an invited talk at *Logic Colloquium '84* in Manchester, but nobody told me about Friedman's paper there.

[3] Parikh had proposed to study such systems already in 1971 [38], but no new results had been published till the late seventies.

[4] The sentences of Paris and Harrington are Π_2, the only "explicit" Π_1 sentences that we know of are Con_T.

an extra uninterpreted predicate (see e.g. [32]). It is still possible that interesting independence results will be obtained by reduction from certain consistency statements, though very little has been done in this direction recently. After all, Paris-Harrington's statement is equivalent to a reflection principle, which is a strengthening of the consistency statement.

2 A brief overview

In this section we shall briefly state the results which we shall discuss below in more detail.

I arrived to the problem of finite consistencies through interpretability of theories. The problem was to characterize interpretability of theories containing a fragment of arithmetic using provability of restricted consistencies. This had been done long before for *reflexive theories* [36, 37, 21]; the main examples of reflexive theories are Peano Arithmetic and Zermello-Fraenkel Set Theory. The property of being reflexive narrows the class of theories too much, e.g. any such theory is not finitely axiomatizable. The solution is to take the restricted consistency statements of Paris and Wilkie [55] and furthermore restrict them to *cuts*. It turns out that these results have other unexpected corollaries, e.g. one can derive the result of Statman mentioned above.

As the condition that the inconsistency proof should lie in a cut is essentially a restriction on the length of such a proof, the next natural step was to investigate the lengths of proofs of finite consistencies $Con_T(n)$ and their versions with restrictions of quantifier complexity. This, in particular, gives more precise bounds in the same applications. Moreover it answers some foundational questions, e.g. proves that the finite modification of Hilbert's program is hardly feasible – for proving $Con_T(n)$ in T one needs a proof of length about n. We have both a lower bound and an upper bound on the lengths of proofs of $Con_T(n)$ in T, and, a bit surprisingly, they are quite close to each other and just around n.

Though these results belong mainly to proof theory, some corollaries can be derived also for the model theory of arithmetic. Namely, the question of how much is the structure of a model M beyond an element a determined by the initial segment $M|_a$ can be solved using knowledge about the lengths of proofs of $Con_T(n)$.

The question whether there is a (say, finitely axiomatized) theory S which proves $Con_T(n)$ for *any* (finitely axiomatized) theory T by a proof of length polynomial in n remains open. So it is possible that this very weak version of Hilbert's program could be realized. It turns out that this question is equivalent to a problem about existence of the most efficient proof system for propositional calculus.

It is, maybe, questionable whether logic can help to solve the persistent deep problems in complexity theory, but if it can, then the following seems to be the most promising way. There are fragments of arithmetic which roughly correspond to various complexity classes. If one could separate (= prove that they are not equal) a pair of such theories, it would be a strong justification for our belief, that the corresponding complexity classes are different. There is essentially only one known method for separating theories which contain some fragment of arithmetic or set theory: Gödel's second incompleteness theorem. Namely the method is based on proving that one theory proves the consistency of the other. This fails for weak fragments of arithmetic, since they are all equiconsistent modulo a very weak theory. Still it is possible that a suitable weakening of the concept of consistency will do. Another application of the bounds on the lengths of proofs of consistency shows that a seemingly natural concept of *bounded consistency* is still too strong for this purpose.

The last result that we will survey is another application of these bounds which gives lower bounds on the lengths of proofs of a little more "mathematical" sentence in case of the theory $I\Delta_0 + \Omega_1$.

3 Basic tools

We shall give some definitions and list a few specific tools on which the proofs are based.

We shall assume in the whole paper that a *theory* is recursively axiomatizable and contains Robinson's arithmetic **Q**. Whenever we talk about the length of a proof in a theory T the particular axiomatization becomes essential. So we rather talk about particular sets of axioms, not just deductively closed sets. In such cases we shall moreover assume that the set of axioms is defined effectively, namely we shall assume that it is a set in $\mathcal{P}$. (Note that $\mathcal{NP}$ would suffice in most cases.)

The Gödel number of a formula φ will be denoted by $\ulcorner\varphi\urcorner$, the length of a formula or a proof will be denoted by $|\ldots|$.

The following is the main lemma which goes back to Gödel [19]. Gödel constructed a concrete selfreferential formula, the general lemma is due to Carnap [8]; in the parametric form it was proved in [34]. (For the proof see e.g. [24].)

Lemma 1 Diagonal Lemma, Fixpoint Theorem. *Let T be a theory and let $\psi(x,y,z)$ be a formula. Then there exists a formula $\varphi(y,z)$ such that*

$$T \vdash \varphi(y,z) \equiv \psi(\ulcorner\varphi(y,\underline{z})\urcorner, y, z),$$

where $\underline{z}$ denotes the z-th numeral, (see below for definition).

Let T be a theory. Then a formula $\varphi(x)$ is called *a cut in* T, if T proves the following sentences:

1. $\varphi(0)$;
2. $\varphi(x) \to \varphi(x+1)$;
3. $\varphi(x) \wedge y \leq x \to \varphi(y)$.

The following are the basic techniques of "cut shortening", which enable to construct cuts with additional properties.

Lemma 2 Solovay, unpublished, see e.g. [24]. *1. For every cut $\varphi(x)$ in T, there exists a cut $\psi(x)$ in T, such that $\psi(x)$ is contained in $\varphi(x)$ and $\psi(x)$ is closed under addition and multiplication (provably in T).*

2. For every cut $\varphi(x)$ in T, there exists a cut $\psi(x)$ in T, such that $\psi(x)$ is contained in $\varphi(x)$ and

$$T \vdash \forall x(\psi(x) \rightarrow \varphi(2^x)),$$

provided the function 2^x is suitably formalized.

Let us note, that in general it is not possible to find a shortening closed under exponentiation (the condition in 2 is weaker).

We shall use the following theories, fragments of Peano Arithmetic: $I\Delta_0$, $I\Delta_0+\Omega$, $I\Delta_0+Exp$, where $I\Delta_0$ is the theory axiomatized by $\mathbf{Q}$ and the induction schema for all bounded formulas, Ω denotes the sentence $\forall x \exists y\ (y = x^{\lceil \log x \rceil})$, and Exp denotes the sentence $\forall x \exists y\ (y = 2^x)$. Since the exponential function is not in the language, we use a Δ_0 definition of the graph of this function (see e.g. [24]).

Here is another important result about cuts.

Theorem 3 Wilkie [55]. *For every cut $\varphi(x)$ in a theory*[5] *T, there exists a cut $\psi(x)$ in T such that $\psi(x)$ is contained in $\varphi(x)$ and such that T proves $I\Delta_0 + \Omega$ restricted to ψ, (i.e. there is an interpretation of $I\Delta_0 + \Omega$ with absolute $+$ and $\cdot$ and the domain contained in φ).*

Furthermore we need a technique of constructing short formulas defined by an iterative schema. Let $\varphi_0(\mathbf{x})$ and $\Phi(X, \mathbf{x})$ be given and suppose we want to construct a sequence of formulas $\varphi_i(\mathbf{x})$, $i = 1, 2, \ldots$, such that

$$\varphi_{i+1}(\mathbf{x}) \equiv \Phi(\varphi_i(\mathbf{x}), \mathbf{x}).$$

Then we can simply substitute $\varphi_0(\mathbf{x})$ in $\Phi(X, \mathbf{x})$ to get $\varphi_1(\mathbf{x})$, then substitute $\varphi_1(\mathbf{x})$ in $\Phi(X, \mathbf{x})$ to get $\varphi_2(\mathbf{x})$ etc. But if X occurs at least twice in $\Phi(X, \mathbf{x})$, the size of such formulas $\varphi_i(\mathbf{x})$ grows exponentially. Ferrante and Rackoff [13] and Solovay [unpublished] have devised techniques which produce formulas $\varphi_i(\mathbf{x})$ of polynomial size.

4 Results

4.1

For a formula $J(x)$ and a sentence φ, we shall denote by φ^J the sentence obtained by restricting the quantifiers to the domain $J(x)$. The following is a strengthening of Gödel's second incompleteness theorem.

Theorem 4 [42] – Theorem 2.1. *For every consistent theory T and every cut $J(x)$ in T, T does not prove Con_T^J.*

[5] Recall that we assume that $\mathbf{Q} \subseteq T$

Put otherwise, there exists a model M of T in which there is a contradiction in T where we, moreover, have the contradiction in J^M. The original proof was an extension of the classical proof of Gödel's Theorem. We shall see that this, as well as the next result, are also easy consequences of bounds on the lengths of proofs of finite consistencies.

A corollary of this result is that Gödel's second incompleteness theorem holds in weak theories, in particular in $\mathbf{Q}$, without any doubts about what Con_T really means there. This is because by Theorem 3 there is a cut in $\mathbf{Q}$ which is a model of $I\Delta_0 + \Omega$. In such a cut all reasonable definitions of Con_T are equivalent.

This result suggests a natural question: what additional conditions can we impose on the proofs of contradiction in Gödel's second incompleteness theorem? The general question has not been answered; the following is an example of a different condition, though also based on restricting the size.

Theorem 5 [42] – Theorem 2.5, [43] – Theorem 4.3. *For every consistent theory T and every Δ_0-formula $I(x)$, there exists a constant k such that, if*

$$T \vdash I(0), I(1), I(2), \ldots,$$

then

$$T + \exists x\ (I(x) \wedge \neg Con_T(x^k))$$

is consistent.

For the following results we have to introduce a restricted provability predicate. The *quantifier complexity* of a formula φ is the maximal number of nesting of quantifiers. *A k-proof* is a proof in which all formulas have quantifier complexity at most k. Note that each first order tautology φ has a cut-free proof and a cut-free proof has the *subformula property.* Hence, if k is at least the quantifier complexity of φ, then φ has a k-proof. Let us note that originally Paris and Wilkie considered the consistency with respect to proofs in which the complexity of every formula is bounded by a constant, where the complexity of terms is disregarded. This definition can also be applied below.

Let $Con_{T,k}$ denote the consistency statement restricted to k-proofs.

A theory is called *sequential*, if there is a formula for coding sequences of elements from the universe which satisfies two axioms: 1. the empty sequence has a code, 2. every sequence s can be prolonged by adding an arbitrary element x. (For a precise definition see [24, page 151].)

The theorem below is a further strengthening of Gödel's second incompleteness theorem, since we use a weaker statement $Con_{T,k}$ instead of Con_T. Moreover the converse shows that it is, in a sense, sharp.

Theorem 6 [42] – Theorem 2.7, Corollary 3.2. *(i) Let T be a finitely axiomatized consistent theory. Then for every cut J in T, there exists a number k such that $T \nvdash Con^J_{T,k}$.*

(ii) Let T be a sequential theory, then for every k, there exists a cut J in T such that $T \vdash Con^J_{T,k}$.

The two statements suggest a natural conjecture: *If T is a finite consistent sequential theory, then there exists a model M of T in which the shortest k-proofs of contradiction, for $k = 1, 2, \ldots$, are cofinal downwards with definable cuts.* This does not follow simply by compactness and requires an additional trick. The conjecture has been proved by Krajíček [27]. For further extensions of these results see Visser [54] and Verbrugge [53]; [54] surveys and gives precise proofs of some previous results whose proofs were only sketched.

The restricted consistency statements can be used to characterize interpretability of a sentence φ in a theory T. An *interpretation* ι of φ in a theory T is given by a translation of the relation and function symbols into formulas of T with the corresponding number of free variables, which inductively defines the translation φ^ι, and by a proof of φ^ι in T. For a more precise definition see [42] or [24]; an interpretation can use parameters etc.

Further concepts that we need are *Herbrand consistency* and *cut-free consistency*. Herbrand's theorem characterizes provability by stating that a sentence φ is provable iff there exists a certain finite object, namely a Herbrand disjunction for φ. We can think of this object as a proof of φ. A theory T is Herbrand consistent, if there is no such proof of the negation of a conjunction of axioms of T. We shall denote Herbrand consistency of T by $HCon_T$. Of course, if a theory S is sufficiently strong, say it is Peano Arithmetic, then $S \vdash HCon_T \equiv Con_T$. So the distinction appears only if we consider weak theories, e.g. $I\Delta_0 + Exp$, or we consider restriction on the lengths of such proofs. The cut-free consistency is a closely related concept. It is based on cut-free proofs in sequent calculi. For more information about this relation see [24, Chapter 5, Section 5].

Theorem 7 [42] – Theorem 3.1. *Let T be a sequential theory and φ a sentence. Then the following are equivalent:*

(i) there is an interpretation of φ in T;

(ii) for every k, there exists a cut J in T such that $T \vdash Con^J_{\varphi,k}$;

(iii) there exists a cut J in T such that $T \vdash HCon^J_\varphi$.

Note that in particular, if T is finitely axiomatized and sequential, then $T \vdash HCon^J_T$ on some cut J in T.

If a theory S has an interpretation in T, then the consistency of T implies the consistency of S, (this is the main application of the concept of interpretability). Does the converse in some modified way hold too? The above theorem gives some reduction to provability of consistencies, but it can hardly be considered as such a converse. This has been achieved in a beautiful theorem of Friedman:[6]

Theorem 8 Friedman [15]. *Let S, T be finite sequential theories. Then the following are equivalent:*

(i) there exists an interpretation of S in T;

[6] This paper also has not been published; the theorem is presented in an article of Smoriński in [46]. Had I known the result, I would not publish my characterization.

(ii) for some m,

$$I\Delta_0 + Exp \vdash \; Con_{T,m} \rightarrow Con_{S,k},$$

where k is the number of quantifiers in S;

(iii)

$$I\Delta_0 + Exp \vdash \; HCon_T \rightarrow HCon_S.$$

Thus S has an interpretation in T iff a certain weak form of relative consistency of S with respect to T is provable in $I\Delta_0 + Exp$.

There is a result of Hájek and Švejdar for Gödel-Bernays set theory GB, which is very much related to Theorem 7. ZF denotes Zermello-Fraenkel set theory, a *set sentence* is a sentence which talks only about sets, i.e. does not talk about classes.

Theorem 9 Hájek and Švejdar [50]. *For every two set sentences φ, ψ, $GB+\varphi$ has an interpretation in $GB+\psi$ iff there exists a cut J in $GB+\psi$ such that*

$$GB + \psi \; \vdash \; Con^J_{ZF+\varphi}.$$

Thus provability in ZF corresponds to Herbrand, or cut-free, provability in GB. There is a result of Friedman in [15] which characterizes interpretability of predicative extensions of theories using relative consistency of the theories, in a way similar to Theorem 8.

We shall give some applications of the results above. First we show that in general cuts cannot be shortened to cuts closed under exponentiation. Originally this was proved by Paris and Dimitracopoulos [40].

Theorem 10 [40], [42] – Corollary 3.7. *There is a cut J in $\mathbf{Q}$ which cannot be shortened to a cut I closed under exponentiation. The same holds for $I\Delta_0$ and $I\Delta_0 + \Omega$.*

Proof. Take a cut J which is a model of $I\Delta_0$ and such that $\mathbf{Q} \vdash HCon^J_{\mathbf{Q}}$. (We do not know if $\mathbf{Q}$ is sequential, but there is a finite theory T which is sequential and which is interpretable in $\mathbf{Q}$.) Suppose there is a cut I in $\mathbf{Q}$ closed under exponentiation. For any fixed k we can transform k-proofs into Herbrand proofs whose size is bounded by a fixed time iterated exponential function. Thus we get all $Con_{\mathbf{Q},k}$ on a single cut I which is impossible by Theorem 6. □

A situation, where in one system the proofs are much shorter than in the other one, is usually called *speed-up*. Next application proves that there is no elementary bound on the speed up of the ordinary proofs with respect to the Herbrand or cut-free proofs. This is a weaker version of Statman's result [49]. We denote by 2^x_k the k-times iterated 2^x.

Theorem 11 [42] – Corollary 3.8. *There is* <u>no</u> *k such that for every n the following holds: if there is a proof of some φ of length $\leq n$, then there is a Herbrand (or cut-free) proof of φ of length $\leq 2^n_k$.*

Proof. Suppose the contrary was true. Let T be a finite sequential fragment of arithmetic in which it is provable. Take a cut J which is a model of $I\Delta_0$ and such that $T \vdash HCon^J_T$. Take I such that

$$T \vdash I(x) \rightarrow J(2^x_k).$$

Then clearly $T \vdash Con^I_T$ which is a contradiction with Theorem 4. □

Now we consider the speed-up of GB over ZF. I want to use this opportunity to correct a mistake in the paper [42]. There I stated the following theorem (Theorem 4.1):

Theorem 12. *There exists a polynomial $p(x)$ such that, for every proof d of a set sentence φ in GB, there exists a proof $d' \leq 2^0_{p(d)}$ of φ in ZF.*

No proof was given there, only an idea of a proof. It has turned out that this idea does not work so easily. Fortunately the theorem is true, such a bound was proved recently by Solovay [48]. Moreover he proved the theorem with a much more precise bound $2^{|\varphi|}_{O(\sqrt{|d|})}$.

Using the above results it is possible to prove a similar nonelementary speed-up of GB over ZF with respect to set sentences [42, Corollary 4.5] (i.e. the common language of the two theories).

Such a speed up holds for other pairs of theories, where one is the predicative extension of the other; e.g. PA and its predicative extension ACA_0. Ignjatović [26] considered the pair PRA (Primitive Recursive Arithmetic) and $I\Sigma_0$. It is a classical result that $I\Sigma_0$ is conservative over PRA, if restricted to the common language, but $I\Sigma_0$ is not a predicative extension of PRA. Though, as Ignjatović proved, there is a similar speed-up of $I\Sigma_0$ over PRA.

4.2

We now turn to the question about the lengths of proofs of finite consistencies $Con_T(n)$. This has a philosophical motivation, as discussed above, but also the bounds on the lengths of proofs of $Con_T(n)$ enable us to get better estimates for speed-up theorems, which we have mentioned above.

We need to introduce some notation. The numeral corresponding to a number n will be denoted by $\underline{n}$ and it is the term

$$a_0 + ((\underline{1} + \underline{1}) \cdot (a_1 + ((\underline{1} + \underline{1}) \cdot (a_2 + \ldots)))),$$

where $a_0, \ldots, a_k \in \{0, 1\}$ and $n = \sum_0^k 2^i a_i$. The syntactical objects are first represented as 0-1 sequences in a natural way and then coded by numbers with a given diadic representation.

The statement that a sentence φ has a proof of length $\leq n$ in a theory T is usually written as $T \vdash^n \varphi$. However it is more convenient to use the notation where $||\varphi||_T$ denotes the length of the shortest proof in T (and ∞ if it is not provable).

Theorem 13 Friedman [14], Pudlák [43, 44]. [7]

(i) Let T be a consistent, $\mathbf{Q} \subseteq T$, T axiomatized by a $\mathcal{P}$ set of axioms. Then for some $\varepsilon > 0$ and every n

$$||Con_T(\underline{n})||_T \geq n^\varepsilon.$$

(ii) If T is sequential and finitely axiomatized, then

$$||Con_T(\underline{n})||_T = O(n).$$

First note that $\underline{n}$, hence also $Con_T(\underline{n})$ has length logarithmic in n, thus the theorem gives nontrivial information. Let us also note that (ii) can be proved also for theories axiomatized by a suitable schema, in particular it holds for PA and ZF, however, the bound is only polynomial, see [43, Theorem 5.5].

The proof of the lower bound is in a sense a finitization of the proof of Gödel's second incompleteness theorem. We take a diagonal sentence $\delta(\underline{n})$ such that

$$T \vdash \delta(\underline{n}) \equiv (||\ulcorner\delta(\underline{n})\urcorner|| > \underline{n})$$

and show that it does not have a proof in T of length $\leq n$. Then we show that $Con_T(\underline{m})$ implies $\delta(\underline{n})$ by a polynomial size proof with m also polynomially bounded by n.

It turns out that the critical part of the proof is the estimate of the length of a proof of the sentence $||\ulcorner\varphi\urcorner||_T \leq \underline{n}$, i.e. how long a proof do we need for proving that φ has a proof of length $\leq n$. The smaller a bound we get, the larger a bound we get for $||Con_T(\underline{n})||_T$, more precisely we get a lower bound which is asymptotically equal to the inverse of the upper bound, see [44, Lemma 2.1].

To see why there is a polynomial upper bound on $||\ulcorner\varphi\urcorner||_T \leq \underline{n}$, we shall introduce the following concept, cf. [30, §1]. We say that $\rho(x_1, \ldots, x_k)$ *$\mathcal{P}$-numerates* a relation $R \subseteq N^k$ in a theory T, if

$$R(n_1, \ldots, n_k) \equiv T \vdash \rho(\underline{n}_1, \ldots, \underline{n}_k),$$

and there exists a Turing machine M running in polynomial time such that M produces a proof of $\rho(\underline{n}_1, \ldots, \underline{n}_k)$ on input $(\underline{n}_1, \ldots, \underline{n}_k)$, whenever $R(n_1, \ldots, n_k)$ is true. Thus, in particular

$$R(n_1, \ldots, n_k) \rightarrow ||\rho(\underline{n}_1, \ldots, \underline{n}_k)||_T \leq p(\log n_1, \ldots, \log n_k),$$

for some polynomial $p(x_1, \ldots, x_k)$. The definition of *polynomial numeration* in [43] is different (it is essentially the *$\mathcal{NP}$-numeration* of [30]) but the theorem below holds also with the stronger concept of $\mathcal{P}$-numeration.

[7] As noted above, Friedman proved only part (i).

Theorem 14 [43] – Theorem 3.2, [30] – Lemma 1.1. *Let T be a consistent, $\mathbf{Q} \subseteq T$, T axiomatized by a $\mathcal{P}$ set of axioms. Then the following are equivalent*
(1) R is in $\mathcal{P}$;
(2) there exists a polynomial numeration of R in T.

The basic property of every proof system is that the proofs can be checked in polynomial time. Also the length of a proof can be computed in polynomial time. Thus we get a polynomial bound on $||\ulcorner\varphi\urcorner||_T \leq \underline{n}$, hence the bound (i) of Theorem 13.

It would be interesting to know the minimal length of a proof of $Con_T(\underline{n})$ exactly, or at least asymptotically. We do not see how to reduce the gap in the ordinary calculus. If we however include an additional rule, we can get a lower bound very close to n. This rule is called *Rule C*. This rule enables to name an object whose existence is proved, i.e. formally

$$\frac{\exists x\ \varphi(x)}{\varphi(c)},$$

where c is a new constant. Using the constant c, we do not have to repeat its property $\varphi(x)$ in the proof, thus the proof becomes shorter. (It is, however, necessary to code the constants efficiently, we cannot use Hilbert's ε-terms.) In this calculus we get a better upper bound on $||\ulcorner\varphi\urcorner||_T \leq \underline{n}$, which gives, see [44, Theorem 2.9],

$$||Con_T(\underline{n})||_T \ = \ \Omega\left(\frac{n}{(\log n)^2}\right),$$

($\Omega(\ldots)$, as usual, denotes a lower bound $\varepsilon \cdot (\ldots)$ for some constant $\varepsilon > 0$).

The upper bound of Theorem 13 is obtained as follows. We define a partial satisfaction relation for formulas of quantifier complexity m. This is done inductively and the technique of Ferrante–Rackoff–Solovay produces such formulas of linear size. To prove their properties we need a proof of size m^2. This gives the following bound:

$$||Con_{T,m}||_T = O(m^2).$$

Furthermore we use:

Lemma 15 [44] – Lemma 3.2. *An optimal size proof of length n contains only formulas of quantifier complexity $O(\sqrt{n})$.*

Thus we need the satisfaction relation only for formulas of quantifier complexity $m = O(\sqrt{n})$, which gives the linear upper bound. □

In order to see that the lower bound on $||Con_T(\underline{n})||_T$ is very useful, let us derive Theorem 4 from Theorem 13(i). We need the following lemma.

Lemma 16 [43] – Lemma 2.2. *Let J be a cut in T, then the sentences $J(\underline{n})$ have proofs of length $p(\log n)$, for some polynomial p.*

To prove the lemma, take a cut I contained in J and closed under plus and times. Since $\underline{n}$ is a term with $O(\log n)$ arithmetic operations, we get a proof of $I(\underline{n})$ of length $p(\log n)$. But J contains I, so we get immediately $J(\underline{n})$.

To finish the proof of Theorem 4, observe that Con_T^J is equivalent to $\forall x(J(x) \rightarrow Con_T(x))$. So if this sentence were provable in T, we would get proofs of $Con_T(\underline{n})$ of length $p(\log n) + O(\log n)$, which is asymptotically less than n^ε. □

The formulas defining partial satisfaction relation have also been used for a proof of *the small reflection principle* in $I\Delta_0 + \Omega$ by Verbrugge [53].

We shall now consider some speed-up consequences of the lower bound on $||Con_T(\underline{n})||_T$. First we observe that we can prove a similar bound for other expressions instead of the numeral $\underline{n}$, we only need to have polynomial proofs that these expressions evaluate to their numerical value. Thus we get

$$||Con_T(2^0_{\underline{n}})||_T \geq (2^0_n)^\varepsilon ,$$

for some $\varepsilon > 0$. Let us note, in passing, that this is an example of an explicit formula with a very long proof. The same proof can be performed with restricted provability, for a sufficiently large k, as the restriction. Thus we get the first part of the following theorem. The second part is proved using cuts, see [44]. ($||\varphi||_{T,k}$ denotes the shortest k-proof of φ in T.)

Theorem 17 [44] – Propositions 4.1 and 4.4. *(i)* $||Con^k_T(2^0_{\underline{n}})||_{T,k} \geq (2^0_n)^\varepsilon$, *where* $\varepsilon > 0$ *and* k *are constants,* k *sufficiently large;*
(ii) $||Con^k_T(2^0_{\underline{n}})||_T = O(n^2)$.

If T is finite and $k = |T|$, then a cut-free proof in T is essentially a k-proof. Thus this result gives a speed-up n to $2^0_{\varepsilon\sqrt{n}}$ for ordinary proofs and cut-free proofs. Up to the ε this is the best possible. This follows from the well-known upper bound $2^n_{O(d)}$ on cut-elimination for a proof of size n and quantifier complexity d, and Lemma 15.

Similar methods give the same speed up n to $2^0_{\varepsilon\sqrt{n}}$ for proving set formulas in GB and ZF. This was suggested in [44] and proved by Solovay [48]; he also showed that this speed up is optimal up to the ε.

4.3

As another application of the lower bound on the lengths of proofs of finite consistency statements, we consider a seemingly unrelated problem from model theory. Let M be a model of a fragment of arithmetic. For an $a \in M$, we denote by $M|_a$ the structure obtained by restricting M to the interval $[0, a]$, where we replace functions by relations. The question is, how much does the structure of $M|_a$ determine the structure of $M|_b$ for some $b \in M$, $b > a$. This means that for a definable function such that $f(a) = b$, we want to know, if there exists a Δ_0 formula $\varphi(x)$ such that $M \models \exists c < f(a)\ \varphi(c)$ while for some other model

$K \not\models \exists c < f(a)\ \varphi(c)$, or vice versa, where we require $M|_a = K|_a$. (Clearly, f must grow faster than any term.)

There is a series of papers on this subject, which gradually decreased the bound f: Paris and Dimitracopoulos [39], Hájek [22], Solovay [47]. We present the bound from [29], which is the best. We shall continue to assume that a theory contains a fragment of arithmetic and it is axiomatized by a $\mathcal{P}$ set of axioms, however, we need a little more than $\mathbf{Q} \subseteq T$, namely $I\Delta_0 + Exp \subseteq T$.

Theorem 18 [29] – Theorem 2.1. *Let M be a countable nonstandard model of T and a, c nonstandard elements of M, $c \leq a$. Then there exists a countable model K of T such that $a \in K$ and*

1. $M|_a = K|_a$;
2. $M|_{2^a} \subseteq K$;
3. $K \models \neg Con_T(a^c)$.

The proof of this theorem uses the lower bound of Theorem 13, it is essential for getting better bounds than the previous ones. Furthermore it uses the Omitting Type Theorem.

Verbrugge [53] sharpened the theorem by proving that assuming $M \models Con_T(a^k)$, for all $k \in \mathbf{N}$, we can find a K with the properties above and for which $M \models Con_T(a^k)$, for all $k \in \mathbf{N}$.

Such results are interesting because they are often connected with open problems in complexity theory. Unfortunately, nobody has succeeded in solving such an open problem using these methods so far. We shall state one corollary of the above theorem which is a result of this type.

Corollary 19 [29] – Proposition 2.6. *Suppose that T proves that every Δ_0-definable set is in $\mathcal{NP}$. Then for every countable model M of T and nonstandard elements $c \leq a \in M$, there exists a model K of T such that $a \in K$ and*

1. $M|_a = K|_a$;
2. $K \models \neg Con_T((\log a)^c)$.

Thus under the assumption that T proves that every Δ_0-definable set is in $\mathcal{NP}$, we can find a K where the difference is much closer to the common segment.[8]

4.4

Theorem 13 settles the problem about the lengths of proofs of $Con_T(\underline{n})$ in T (up to a polynomial gap). If S is essentially stronger than T, namely if $S \vdash Con_T$, then S proves $Con_T(\underline{n})$ by a proof of length $O(\log n)$ (just derive $Con_T(\underline{n})$ from $\forall x\ Con_T(x)$ by substitution). However, if S is *weaker* than T, the problem

[8] Note that proofs of length $(\log a)^c$ have Gödel numbers of size about $K^{(\log a)^c}$, K a constant.

remains open. A natural conjecture is e.g. that $||Con_T(\underline{n})||_S$ is exponential, if $T \vdash Con_S$. From the foundational point of view the most interesting question is if there is an S which proves all sentences $Con_T(\underline{n})$ by polynomial size proofs. These problems turn out to be closely related to open problems in complexity theory and propositional calculus; we are going to consider these relations in this section.

A proof system for propositional calculus P is any sound and complete system where the proofs can be checked in polynomial time. Formally it means that we have a polynomial time Turing machine M which accepts an input (t, d) only if t is a propositional tautology and for every propositional tautology t there exists a string d such that M accepts the input (t, d). The question whether there exists a propositional proof system in which all tautologies have polynomial size proofs is equivalent to the open problem $\mathcal{NP} = co\mathcal{NP}$?. Our present knowledge does not rule out that such a system is just the usual proof system based on a finite number of axiom schemas and *Modus Ponens*. We introduce a quasiordering relation on propositional proof systems by defining

> $P \leq Q$ if there exists a polynomial p such that for every tautology t, if. t has a proof of length n in Q, then t has a proof of length $\leq p(n)$ in P.

In words it means that, up to a polynomial increase, the system P is at least as efficient as the system Q. Even if $\mathcal{NP} \neq co\mathcal{NP}$ it is possible that there exists a least element in the quasiordering.

To simplify the matter, we shall assume throughout this section that *all theories are finite*. Let us first note that if $S \vdash \mathcal{NP} = co\mathcal{NP}$, (it suffices $S \vdash \mathcal{NEXP} = co\mathcal{NEXP}$), then S proves all sentences $Con_T(\underline{n})$ by polynomial size proofs. Roughly speaking this is because the condition of "checking all proofs of a given length" can be replaced by "checking a single one".

A set of strings X is called *sparse*, if for every length n, there are only polynomially many elements of this length in X.

The following theorem relates important problems from three different fields.

Theorem 20 [30] – Theorem 2.1. *The following are equivalent:*

(1) There exists a finite fragment S of arithmetic such that for every finitely axiomatized consistent theory T there exists a polynomial p such that $||Con_T(\underline{n})||_S \leq p(n)$, for every $n \in N$.

(2) There exists a propositional proof system which is a least element in the quasiordering (i.e. which is the most efficient with respect to the lengths of proofs).

(3) For every set $A \in co\mathcal{NP}$ there exists a nondeterministic Turing machine M which accepts A and such that for every $X \subseteq A$, $X \in \mathcal{P}$, X sparse, M accepts all elements of X in polynomial time (i.e. M is efficient on all "easy" subsets).

To give an idea about the proof, we shall show a construction of a propositional proof system from a theory and a construction of a theory from a propositional proof system. Let $Taut(x)$ be a suitable formalization of the concept of a propositional tautology, (namely we take the formula $\forall y\ Sat(x,y)$, where $Sat(t,a)$ is a $\mathcal{P}$-numeration of the predicate t *satisfied by an assignment* a).

Let T be a theory. Then we define a propositional proof system $P(T)$ by saying that d is a proof of tautology t, iff d is a proof of the sentence $Taut(\ulcorner t \urcorner)$ in T.

Now, let P be a propositional proof system. Let $Prf(t,d)$ be a $\mathcal{P}$-numeration of the proof predicate of P. Then we construct a theory $S(P)$ by taking a suitable basic fragment of arithmetic T_0 and adding the reflection principle

$$\forall x, y(Prf(x,y) \rightarrow Taut(x)).$$

Furthermore we need to translate some first order sentences into propositional formulas. Thus in particular $Con_T(\underline{n})$ can be translated onto a propositional tautology $t_{Con_T(\underline{n})}$ whose length is polynomial in n.

Very little progress has been made in solving problems in complexity theory such as $\mathcal{NP} = co\mathcal{NP}$? or the problem in Theorem 20. The only successful approach is based on showing that the problems cannot be solved using certain restricted means. The idea is to use the so called *relativized problems*, which means that we take a suitable set A, called *an oracle*, and allow the Turing machines mentioned in the problems to use information about this set for free. Thus we obtain a relativized class $\mathcal{X}^A$ for every complexity class $\mathcal{X}$. E.g. one can construct A and B such that $\mathcal{NP}^A = co\mathcal{NP}^A$ and $\mathcal{NP}^B \neq co\mathcal{NP}^B$ which proves that this problem cannot be solved using techniques which relativize. In [30, Theorem 4.1] we show that there is an oracle C such that the statement (3) of Theorem 20 is false, if relativized to C. Together with the relativization $\mathcal{NP}^A = co\mathcal{NP}^A$ this gives a kind of independence for the problems in Theorem 20.

Also using relativizations, Verbickij [52] showed that (3) is weaker than $\mathcal{NEXP} = co\mathcal{NEXP}$, thus it seem really hard to refute the statements in Theorem 20.

4.5

We consider two more applications of the lower bound on finite consistency statements.

The first one concerns the problem of separation of fragments of arithmetic. Buss [6] introduced bounded arithmetic S_2 and its fragments S_2^i. The language of these theories contains the usual language of arithmetic plus the function symbols $\lfloor 1/2x \rfloor$, $|x|$, and $x\#y$ whose interpretation is

$$|x| = \lceil \log_2(x+1) \rceil \text{ and } x\#y = 2^{|x|\cdot|y|}.$$

S_2 is axiomatized by a finite set of basic axioms and the following schema of induction

$$\varphi(0) \wedge \forall x(\varphi(\lfloor 1/2x \rfloor) \rightarrow \varphi(x)) \rightarrow \forall \varphi(x),$$

for all bounded formulas. This theory is a conservative extension of $I\Delta_0+\Omega$. The fragments S_2^i are defined by restricting the schema of induction to Σ_i^b-formulas, where the classes Σ_i^b are defined so that the formulas define just the sets in the Polynomial Hierarchy Σ_i^p. In particular Σ_1^b defines the $\mathcal{NP}$ sets, so the theory S_2^1 is closely related to the class $\mathcal{NP}$.

Similarly as it is an open problem if the Polynomial Hierarchy is increasing, it is an open problem if the hierarchy of theories S_2^i is increasing. The subtheories of Peano Arithmetic $I\Sigma_i$ can be separated by sentences $Con_{I\Sigma_i}$ which are provable in $I\Sigma_{i+1}$, but not in $I\Sigma_i$. This is not possible for S_2^i's, since the whole S_2 does not prove even $Con_{\mathbf{Q}}$. In fact it is even worse:

Theorem 21 Wilkie and Paris [56].

$$I\Delta_0 + Exp \nvdash Con_{\mathbf{Q}}.$$

A natural modification of the concept of consistency has been considered, where one talks only about proofs of contradiction which use only *bounded formulas.* Such a restricted statement of consistency of a theory T is denoted by $BDCon_T$. Though it looks like the right concept for Bounded Arithmetic, it turns out that it is still too strong for such a separation:

Theorem 22 [45].

$$S_2 \nvdash BDCon_{S_2^1}.$$

Thus we cannot separate in such a way even S_2 from S_2^1. This improves a former result of Buss [6], where he showed that $S_2^{i+1} \vdash BDCon_{S_2^i}$ holds for at most one i. The result has been further improved by Takeuti [51] and Buss and Ignjatović [7]. Let us note that Krajíček and Pudlák [31], and Krajíček and Takeuti [33] found sentences that separate such fragments, if the fragments are different, however, since these sentences are rather "combinatorial" than "logical", it seems impossible to use diagonalization arguments for them.

The proof of the theorem is based on a modification of the lower bound on the lengths of proofs of $Con_T(\underline{n})$, in which we consider the Gödel numbers of the proofs instead of their lengths. Let $\gamma(\varphi)$ denote the minimal Gödel number of a proof of the sentence φ in S_2^1. Let $BDCon^{\#}_{S_2^1}(x)$ denote that there is no proof of contradiction in S_1^2 whose Gödel number is $\leq x$; it is a bounded formula in the language of S_2. Using a similar proof as above, we get:

Lemma 23 [45] – Lemma 1. *For every term $s(x)$, there exists a term $r(x)$ such that for all but finitely many n*

$$\gamma\left(BDCon^{\#}_{S_2^1}(r(\underline{n}))\right) > s(n).$$

The proof of Theorem 22 uses another lemma, which states that for every bounded formula φ

$$\gamma(\varphi(\underline{n})) \leq p(n\#n),$$

for some polynomial p, if $S_2 \vdash \forall x\ \varphi(x)$. The proof is finished by taking $s(x) = x\#x\#x$, which grows faster than $p(n\#n)$ for any polynomial p, and $\varphi(x) = BDCon^{\#}_{S^1_2}(x)$. □

The last application is to the speed-up obtained by *proving the provability of* φ instead of proving φ. Parikh [38] has shown that for Peano arithmetic this speed-up cannot be bounded by any recursive function. We observe that we can get concrete examples for such a speed up from consistency statements. E.g. we have the following upper bound:

Proposition 24 [23] – Theorem 1,(b). *Let $Prov_T$ be a suitable formalization of the provability in T. Then*

$$||Prov_T\left(\ulcorner Con_T(2^0_{\underline{n}})\urcorner\right)||_T \ \leq\ p(n),$$

for some polynomial p.

With the bound

$$||Con_T(2^0_{\underline{n}})||_T \ \geq\ \left(2^0_{\underline{n}}\right)^{\varepsilon},$$

$\varepsilon > 0$, which we have already considered, we get a more than elementary speed-up for an explicit sequence of sentences. It would be interesting to prove such speed-ups for more natural mathematical sentences. Presently this can be done only for sentences expressing existence of fast growing functions. We conclude our survey with such a result.

Theorem 25 [23] – Theorem 3. *(a) There exists a polynomial p such that for all n*

$$||Prov_{I\Delta_0+\Omega}\left(\ulcorner\exists y(y = 2^0_{2^0_{\underline{n}}})\urcorner\right)||_{I\Delta_0+\Omega} \ \leq\ p(n).$$

(b) For every sufficiently large n,

$$||\exists y(y = 2^0_{2^0_{\underline{n}}})||_{I\Delta_0+\Omega} \ \geq 2^0_n/4.$$

5 Conclusions

We have shown that bounds on finite consistency statements are useful for several purposes. They can be used to show speed-up for different logical calculi, speed-up between some theories and speed-up when we prove the existence of proofs instead of constructing them. They are also explicit examples of formulas with long proofs.

The problem about the lengths of proofs of $Con_T(\underline{n})$ has been solved only for T itself and stronger theories.[9] If S is weaker than T, then $||Con_T(\underline{n})||_S$ can

[9] To be quite precise, if S proves Con_T, then, trivially $Con_T(\underline{n})$ has proofs of length $O(\log n)$, for $S = T$ it is about n, but for S stronger than T which does not prove Con_T, we do not know; we only know that it is between these values.

be exponential in n. I think this is a fundamental problem. The connection with problems in complexity theory shows that it is a very difficult problem.

Suppose we proved that for every S, there exists T such that $||Con_T(\underline{n})||_S$ is exponential. Then this would solve a lot of problems in complexity theory, in particular it would imply $\mathcal{P} \neq \mathcal{NP}$. It does not seem to be possible to use some higher order argument from logic to solve the problems in complexity theory, without really going into deep combinatorial problems. But still, maybe we are just missing some simple clever idea. I imagine that Hilbert had similar feelings about the consistency problem before Gödel solved it.[10] The problem of consistency is a statement about application of finite rules finitely many times, so why anything else than finite combinatorics is needed? Though Gödel had to do some technical work, his basic trick was the selfreference (essentially the paradox of liar, which was known already to ancient Greeks), which is a higher order argument which has little to do with finite combinatorics.

Let us conclude by repeating a conjecture that we stated in [42]:

Conjecture *If T is sufficiently strong, then $||Con_{T+Con_T}(\underline{n})||_T$ is not bounded by a polynomial.*

The intuition is that $T + Con_T$ is essentially stronger than T, since it proves $Con_T(\underline{n})$ by proofs of logarithmic length, while T needs n^ϵ, therefore T should not have short proofs of $Con_{T+Con_T}(\underline{n})$. As stated above, the positive answer would solve fundamental problems in complexity theory.

Maybe the solution of $\mathcal{P} = \mathcal{NP}$? etc. will be as surprising as was Gödel's ...

References

1. Miklos Ajtai. The complexity of the pigeonhole principle. In *Proceedings of the 29th Annual IEEE Symposium on Foundations of Computer Science*, pages 346–355, 1988.
2. Miklos Ajtai. The independence of the modulo p counting principles. In *Proc. 26th ACM Symp. on Theory of Computing*, pages 402–411, 1994.
3. Mathias Baaz and Pavel Pudlák. Kreisel's conjecture for $L\exists_1$. In Peter Clote and Jan Krajíček, editors, *Arithmetic Proof Theory and Computational Complexity*, pages 30–39. Oxford Univ. Press, 1993.
4. Paul Beame, Russell Impagliazzo, Jan Krajíček, Toniann Pitassi, Pavel Pudlák, and Alan Woods. Exponential lower bounds for the pigeonhole principle. In *Proceedings of the 24th Annual ACM Symposium on Theory of Computing*, pages 200–221, 1992.
5. A. Bezboruah and J. Shepherdson. Gödel's second incompleteness theorem for Q. *J. Symbolic Logic*, pages 503–512, 1976.

[10] Hilbert hoped that his consistency program will be realized using the new theory that he invented and named *Beweistheorie* [25] (nowadays his theory is called *ϵ-calculus*).

6. Samuel R. Buss. *Bounded Arithmetic.* Bibliopolis, 1986. Revision of 1985 Princeton University Ph.D. thesis.
7. Samuel R. Buss and Aleksandar Ignjatović. Unprovability of consistency statements in fragments of bounded arithmetic. In preparation, 1993.
8. R. Carnap. *Logische Syntax der Sprache.* Springer-Verlag, 1934.
9. Stephen A. Cook. Feasibly constructive proofs and the propositional calculus. In *Proceedings of the 7th Annual ACM Symposium on Theory of Computing,* pages 83–97, 1975.
10. Martin Dowd. *Propositional Representation of Arithmetic Proofs.* PhD thesis, University of Toronto, 1979.
11. A.G. Dragalin. Correctness of inconsistent theories with notions of feasibility. volume 108 of *Lecture Notes in Comp. Sci.,* pages 58–79, 1985.
12. A. Esenine-Volpin. Le programme ultra-intuitionniste des fondements des mathematiques. In *Infinitistic Methods, Proceedings of the Symposium on Foundations of Mathematics,* pages 201–223. PWN, Warsaw, 1961.
13. J. Ferrante and Ch. W. Rackoff. *The Computational Complexity of Logical Theories.* LNM 718. Springer-Verlag, 1979.
14. H. Friedman. On the consistency, completeness, and correctness problems. Ohio State Univ., unpublished, 1979.
15. H. Friedman. Translatability and relative consistency II. Ohio State Univ., unpublished, 1979.
16. Yu.V. Gavrilenko. Monotone theories of feasible numbers. *Doklady Akademii Nauk SSSR,* 276(1):18–22, 1984.
17. Gerhard Gentzen. Die Widerspruchsfreiheit der reinen Zahlentheorie. *Mathematische Annalen,* 112:493–565, 1936. English translation in *Gerhard Gentzen, Collected Papers of Gerhard Gentzen,* Editted by M. E. Szabo, North-Holland, 1969, pp. 132-213.
18. Gerhard Gentzen. *Collected Papers of Gerhard Gentzen.* North-Holland, 1969. Editted by M. E. Szabo.
19. K. Gödel. Über formal unentscheidbare Sätze der Principia Mathematica und vewandter Systeme I. *Monatshefte Math. Phys.,* 38:173–198, 1931.
20. K. Gödel. Über die Länge von Beweisen. *Ergebnisse eines Mathematischen Kolloquiums,* pages 23–24, 1936. English translation in *Kurt Gödel: Collected Works, Volume 1,* pages 396-399, Oxford University Press, 1986.
21. P. Hájek. On interpretability in set theories II. *Commentatione Math. Univ. Carol.,* 13:445–455, 1972.
22. Petr Hájek. On a new notion of partial conservativity. In *Logic Colloquium '83,* pages 217–232. Springer-Verlag, 1983.
23. Petr Hájek, Franco Montagna, and Pavel Pudlák. Abbreviating proofs using metamathematical rules. In Peter Clote and Jan Krajíček, editors, *Arithmetic Proof Theory and Computational Complexity,* pages 197–221. Oxford Univ. Press, 1993.
24. Petr Hájek and Pavel Pudlák. *Metamathematics of First-order Arithmetic.* Springer-Verlag, 1993.
25. D. Hilbert. *Die Grundlagen der Mathematik,* volume 5 of *Hamburger Mathematische Einzelschriften.* Teubner, 1934. A lecture presented in Hamburg in July 1927.
26. A. Ignjatović. *Fragments of First and Second Order Arithmetic and Length of Proofs.* PhD thesis, University of California at Berkeley, 1990.
27. Jan Krajíček. A note on proofs of falsehood. *Archiv f. Math. Logic u. Grundlagen d. Math.,* 26:169–179, 1987.

28. Jan Krajíček. Lower bounds to the size of constant-depth propositional proofs. *Journal of Symbolic Logic*, 59(1):73–86, 1994.
29. Jan Krajíček and Pavel Pudlák. On the structure of initial segments of models of arithmetic. *Archive for Mathematical Logic*, 28:91–98, 1989.
30. Jan Krajíček and Pavel Pudlák. Propositional proof systems, the consistency of first-order theories and the complexity of computations. *Journal of Symbolic Logic*, 54:1063–1079, 1989.
31. Jan Krajíček and Pavel Pudlák. Quantified propositional calculi and fragments of bounded arithmetic. *Zeitschrift für Mathematische Logik und Grundlagen der Mathematik*, 36:29–46, 1990.
32. Jan Krajíček, Pavel Pudlák, and Gaisi Takeuti. Bounded arithmetic and the polynomial hierarchy. *Annals of Pure and Applied Logic*, 52:143–154, 1991.
33. Jan Krajíček and Gaisi Takeuti. On induction-free provability. *Annals of Math. and Artificial Intelligence*, 6:107–126, 1992.
34. R. Montague. Theories incomparable with respect to relative interpretability. *Journal of Symbolic Logic*, 27:195–211, 1962.
35. V. P. Orevkov. Correctness of short proofs in theory with notions of feasibility. volume 417 of *Lecture Notes in Comp. Sci.*, pages 242–245, 1990.
36. S. Orey. Relative interpretations. *Journal of Symbolic Logic*, 24:281–282, 1959.
37. S. Orey. Relative interpretations. *Zeitschrift für Mathematische Logik und Grundlagen der Mathematik*, 7:146–153, 1961.
38. R. Parikh. Existence and feasibility in arithmetic. *J. Symbolic Logic*, 36:494–508, 1971.
39. J. B. Paris and C. Dimitracopoulos. Truth definitions for formulas. In *Logic et algoritmic, L'enscignment Mathematique No 30*, pages 318–329, 1982.
40. J. B. Paris and C. Dimitracopoulos. A note on the undefinability of cuts. *J. Symbolic Logic*, 48:564–569, 1983.
41. J.B. Paris and L. Harrington. A mathematical incompleteness in Peano Arithmetic. In *Handbook of Mathematical Logic*, pages 1133–1142. North-Holland, 1977.
42. Pavel Pudlák. Cuts, consistency statements and interpretation. *Journal of Symbolic Logic*, 50:423–441, 1985.
43. Pavel Pudlák. On the lengths of proofs of finitistic consistency statements in first order theories. In *Logic Colloquium '84*, pages 165–196. North-Holland, 1986.
44. Pavel Pudlák. Improved bounds to the lengths of proofs of finitistic consistency statements. In *Logic and Combinatorics*, volume 65 of *Contemporary Mathematics*, pages 309–331. American Mathematical Society, 1987.
45. Pavel Pudlák. A note on bounded arithmetic. *Fundamenta Mathematicae*, 136:85–89, 1990.
46. C. Smoriński. Nonstandard models and related developments. In *Harvey Friedman's Research on the Foundations of Mathematics*, pages 179–229. North Holland, 1985.
47. R.M. Solovay. Injecting inconsistencies into models of PA. *Annals of Pure and Applied Logic*, 44:101–132, 1989.
48. R.M. Solovay. Upper bounds on the speedup of GB over ZF. preprint, 1990.
49. R. Statman. Proof search and speed-up in the predicate calculus. *Ann. Math. Logic*, 15:225–287, 1978.
50. V. Švejdar. Modal analysis of generalized Rosser sentences. *Journ. of Symb. Logic*, 48:986–999, 1983.
51. G. Takeuti. Some relations among systems for bounded arithmetic. *Annals of Pure and Applied Logic*, 39:75–104, 1988.

52. O.V. Verbitsky. Optimal algorithms for coNP-sets and the problem EXP=?NEXP. *Matematicheskie zametki*, 50(2):37–46, 1991. Eglish translation in: Math. Notes 50,1-2, (1991), pp. 798-801.
53. L.Ch. Verbrugge. *Efficient Mathematics.* PhD thesis, Universiteit van Amsterdam, 1993.
54. A. Visser. The unprovability of small consistency. *Archive for Math. Logic*, 32:275–298, 1993.
55. A. Wilkie. On sentences interpretable in systems of arithmetic. In *Logic Colloquium '84*, pages 329–342. North-Holland, 1986.
56. A.J. Wilkie and J.B. Paris. On the schema of induction for bounded arithmetical formulas. *Annals of Pure and Applied Logic*, 35:261–302, 1987.

The Craig Interpolation Theorem for Schematic Systems

A. CARBONE *

Institut für Algebra und Diskrete Mathematik
Technische Universität Wien
e-mail: ale@logic.tuwien.ac.at

Abstract. The notion of Schematic System has been introduced by Parikh in the early seventies. It is a metamathematical notion describing the concept of deduction system and the operation of substitution of terms and formulas in it. We show a generalization of the Craig Interpolation Theorem for a natural class of schematic systems while we determine sufficient conditions for a schematic system to enjoy Interpolation. These conditions are much weaker than the usual conditions of symmetricity that are satisfied by the logics usually studied. The proof of the Craig Interpolation Theorem that we propose is a refinement of Maehara's construction and it is based on the idea of tracing the flow of formulas in a proof.

1 Introduction

For two given formulas A and B such that $A \to B$ is provable, there is a formula C (called *interpolant*) expressed in the language common to A, B (roughly all symbols in C appear in both A and B) and such that $A \to C$ and $C \to B$ are provable.

This is the statement of the Interpolation Theorem proved for classical logic by Craig in 1957 ([5]). Later, several authors showed that the statement holds indeed for many other logical systems such as intuitionistic logic, infinitary logics, certain multiple-valued logics, fragments of Linear Logic, and so forth. We will prove a generalization of Craig's Theorem for a class of logical systems (called *regular schematic systems*) and show that the theorem is properly a statement about the structure of derivations and it is not based on the validity of the calculus. For instance the cut-free system of classical logic extended with a pair of rules of the form

$$\frac{\Gamma \to \Delta, P}{\Gamma \to \Delta, P \wedge \bot} \qquad \frac{P, \Gamma \to \Delta}{P \wedge \bot, \Gamma \to \Delta}$$

(with Γ, Δ being two arbitrary collections of formulas) enjoys the Interpolation Property even though the left rule is *not* valid.

* Partially supported by the grant # ERB4001GT940693 from the EC (HCM Program) and the Lise-Meitner Stipendium # M00187-MAT (Austrian FWF.)

We will see that the logical form of the rules allowed in regular schematic systems does not require the well-known *subformula property* (i.e. roughly speaking all formulas occurring in the antecedent of a rule should appear as (sub)formulas in the consequent) to hold. It will just require a weaker 'embedding' condition (as discussed in section 2).

The class of logical systems we consider will be defined through the metamathematical notion of *schematic system* introduced by Parikh in the early seventies. Section 2 will be devoted to a discussion of schematic systems and regular schematic systems. In section 3 we introduce logical graphs tracing the flow of occurrences of formulas in a proof. Our definition is a variant of the notion of *logical flow graph* introduced in [2]. Similar notions appeared also in [7]. In section 4 we use the graph theoretic tool to prove Craig's Theorem for regular schematic systems. As a corollary of the result, we derive the interpolation property for several systems known in the literature. Section 5 contains a discussion about further development of this metamathematical approach to logical systems.

The author wishes to thank Matthias Baaz and Rohit Parikh for comments on an earlier version of this manuscript, and Stephen Semmes for stimulating conversations on the topic and helpful remarks.

2 Schematic Systems

Axiom schemes and rules of inference are generally explained in the literature with the help of formula variables, so in order to define the notion of *schematic system* one can do the obvious, namely, expand the notation of the predicate calculus to include metamathematical symbols and emphasize *substitution* of terms and formulas as the central idea. The notion of schematic system was first introduced by Parikh in [10]. The notion has been used in further work in [9], [6].

We will use Gentzen style sequent schematic systems where we can easily formalize proofs as trees of sequents. We will think of each sequent as used only once in the proof.

2.1 The language

To introduce the concepts we follow Parikh's presentation ([10]). The language $\mathcal{L}$ will include

1. variables $x, y, z, \ldots$;
2. constants $c, c_1, c_2, \ldots$;
3. first order function symbols $f, g, \ldots$;
4. first order n-ary (for $n > 0$) predicate constants $F(x), G(x, y), \ldots$ (possibly the binary equality symbol $=$);
5. special predicate constants $\top, \bot, \ldots$;

6. logical symbols $\wedge, \vee, \supset, \neg, \exists, \forall, \ldots$;
7. other symbols $(,), [,], \ldots$.

We do not assume that the whole apparatus mentioned above is present in our language; on the other hand, we allow *extra* symbols to occur, for instance new types of connectives or higher order variables [2].

The language $\mathcal{L}$ includes as metamathematical symbols

1. metavariables $u, v, w, u_1, \ldots$;
2. term variables $r, s, t, r_1, \ldots$;
3. n-ary predicate (formula) variables (for $n \geq 0$) $P, Q(u), R(u, v), \ldots$.

Terms will be formed from variables, constants, metavariables and term variables, by means of function symbols. Terms not containing metanotation will be called *regular terms.* Atomic formulas will be formed from predicate variables and predicate constants by suitably placing terms as their arguments. Formulas will be formed from atomic formulas, special predicate constants, connectives and quantifiers in the usual way. Formulas not containing metanotation will be called *regular formulas.* Terms and formulas will be denoted by Greek and script letters respectively, Roman if they are regular.

Since we will work with a Gentzen style formulation of a calculus, we say that a *schematic sequent* is a sequent $\mathcal{A}_1, \ldots, \mathcal{A}_l \rightarrow \mathcal{A}_{l+1}, \ldots, \mathcal{A}_k$ of formulas in $\mathcal{L}$. The shorthand notation $\mathcal{S}[\mathcal{A}_1, \ldots, \mathcal{A}_n]$ denotes a schematic sequent containing the formula occurrences $\mathcal{A}_1, \ldots, \mathcal{A}_n$ and maybe other formulas as well. The position (i.e. left or right) of the formulas $\mathcal{A}_1, \ldots, \mathcal{A}_n$ in $\mathcal{S}[\mathcal{A}_1, \ldots, \mathcal{A}_n]$ is intended to be fixed. Namely, everytime we write $\mathcal{S}[\mathcal{A}_1, \ldots, \mathcal{A}_n]$ we really mean to write $\mathcal{S}[\mathcal{A}_1^{i_1}, \ldots, \mathcal{A}_n^{i_n}]$ (with $i_j \in \{1, 2\}$ and $j = 1 \ldots n$) where $\mathcal{A}_j^1$ denotes the formula A_j occurring in the left hand side of the sequent arrow, and $\mathcal{A}_j^2$ denotes A_j occurring in the right hand side.

Although we will not analyse specific logical systems here, let us describe briefly how a specific system can be derived by substitution from a schematic system. A *substitution* δ will be an assignment of variables $x_1, \ldots, x_n$ to certain metavariables $u_1, \ldots u_n$, of regular terms $\alpha_1, \ldots, \alpha_m$ to term variables $t_1, \ldots, t_m$ and of regular formulas $A_1, A_2, \ldots$ to certain predicate variables $P, Q(u), \ldots$. Substitutions will be informal objects used to study formal systems and variables will always be such as to avoid conflicts.

The substitution δ induces a map that we will also call δ, from certain terms to regular terms uniquely defined by

$$\delta(u_i) = x_i, \quad \delta(t_j) = \alpha_j, \qquad i = 1 \ldots, n, \quad j = 1 \ldots m$$

and

$$\delta(f(\sigma_1, \ldots \sigma_k)) = f(\delta(\sigma_1), \ldots, \delta(\sigma_k))$$

[2] Variables of order $n + 1$ are intended to range over the set of objects of order n; in particular, variables have no arguments and the only atomic expression in which $(n+1)$-th order variables $\mathcal{X}$ can occur is $X \in \mathcal{X}$ where X is of order n.

provided that the right hand side is defined and that k is the arity of the function symbol f.

The substitution δ also induces a map on formulas $\mathcal{F}$ defined by

1. $\delta(\mathcal{F}) = F(\delta(\sigma_1), \ldots, \delta(\sigma_k))$ when $\mathcal{F}$ is the atomic formula $F(\sigma_1, \ldots, \sigma_k)$ with predicate constant F and the right hand side is defined;
2. $\delta(\mathcal{F}) = A[\delta(\sigma_1), \ldots, \delta(\sigma_k)]$ when $\mathcal{F}$ is the atomic formula $P(\sigma_1, \ldots, \sigma_k)$ with predicate variable P such that $\delta(P(u_1, \ldots u_k)) = A[x_1, \ldots x_n]$ (where the symbol $A[x_1, \ldots x_n]$ means that A is a regular formula containing the variables $x_1, \ldots x_n$) and the right hand side is defined;
3. $\delta(\mathcal{F} \wedge \mathcal{G}) = \delta(\mathcal{F}) \wedge \delta(\mathcal{G})$, etc.;
4. $\delta((\forall x)\mathcal{F}) = (\forall x)\delta(\mathcal{F})$, etc.;
5. $\delta((\forall u)\mathcal{F}) = (\forall \delta(u))\delta(\mathcal{F})$, etc.

Finally a substitution δ applied to a sequent $\mathcal{A}_1, \ldots, \mathcal{A}_l \rightarrow \mathcal{A}_{l+1}, \ldots, \mathcal{A}_k$ is defined as the sequent $\delta(\mathcal{A}_1), \ldots, \delta(\mathcal{A}_l) \rightarrow \delta(\mathcal{A}_{l+1}), \ldots, \delta(\mathcal{A}_k)$.

A *restriction* is called *admissible* if it is of the form 'provided σ is free for u (for x) in P ($\mathcal{S}$)' or 'provided u (x) is not free in P ($\mathcal{S}$)' or 'provided u (x) does not occur in P ($\sigma, \mathcal{S}$)' where u is a metavariable, x is a variable, σ is a term, P a predicate variable and $\mathcal{S}$ is a schematic sequent.

It is clear what we intend by saying that a substitution δ obeys a restriction R, e.g. δ obeys 'provided σ is free for u in P' iff $\delta(\sigma)$ is free for $\delta(u)$ in $\delta(P)$. Whenever we will refer to restrictions we will intend them to be admissible restrictions.

We are now ready to define what are a schematic axiom and a schematic rule. A *schematic axiom* is a pair $(\mathcal{S}, R)$ where R is a finite set of restrictions and $\mathcal{S}$ is a schematic sequent of the form

$$\mathcal{A}_1, \Gamma \rightarrow \Delta, \mathcal{A}_2$$

where $\mathcal{A}_1$ and $\mathcal{A}_2$ are called *distinguished* occurrences in the axiom and Γ, Δ denote arbitrary collections (i.e. sets that possibly contain multiple occurrences of the same formula) of formulas called *side* formulas of the axiom. We ask either that all (non special) predicate constants and predicate variables occurring in $\mathcal{A}_1$ also occur in $\mathcal{A}_2$ or the other way around. We allow a schematic axiom to have no distinguished occurrences in one of the sides of the sequent arrow. A schematic axiom has to have at least one distinguished occurrence though. We assume that sequents of the form $\mathcal{A}, \Gamma \rightarrow \Delta, \mathcal{A}$, $\Gamma \rightarrow \Delta, \top$ and $\bot, \Gamma \rightarrow \Delta$ are schematic axioms.

To avoid any confusion on side formulas, let us remark that two schematic axioms $\mathcal{A}_1, \Gamma \rightarrow \Delta, \mathcal{A}_2$ and $\mathcal{A}_1, \Gamma' \rightarrow \Delta', \mathcal{A}_2$ are assumed to be the same schematic axiom. An axiom $A_1, \Theta \rightarrow \Lambda, A_2$ is *derived* from a schematic axiom $\mathcal{A}_1, \Gamma \rightarrow \Delta, \mathcal{A}_2$ if there is a substitution map δ such that $\delta(\mathcal{A}_1) = A_1$ and $\delta(\mathcal{A}_2) = A_2$, and Θ, Λ are regular formulas.

Similarly, a *schematic rule* of *arity* k is a sequence

$$(\mathcal{S}_0[\mathcal{A}_0], \mathcal{S}_1[\mathcal{A}_{1,1}, \ldots, \mathcal{A}_{1,n_1}], \ldots, \mathcal{S}_k[\mathcal{A}_{k,1}, \ldots, \mathcal{A}_{k,n_k}], R)$$

where R is a finite set of restrictions and $\mathcal{S}_0, \mathcal{S}_1, \ldots, \mathcal{S}_k$ are schematic sequents. A schematic rule will be usually written as

$$\frac{\mathcal{S}_1[\mathcal{A}_{1,1}, \ldots, \mathcal{A}_{1,n_1}], \ldots, \mathcal{S}_k[\mathcal{A}_{k,1}, \ldots, \mathcal{A}_{k,n_k}]}{\mathcal{S}_0[\mathcal{A}_0]}$$

where $\mathcal{A}_{i,1}, \ldots, \mathcal{A}_{i,n_i}$ are the *auxiliary* occurrences in the i-th sequent of the rule and $\mathcal{A}_0$ is the *main* formula of the rule. All formulas occurring in the sequents $\mathcal{S}_i$ which are neither an auxiliary formula (i.e. $A_{i,j}$, for some $i = 1 \ldots k$, $j = 1 \ldots n_i$) nor a main formula (i.e. A_0), are called *side* formulas of the rule. If the condition 'provided u (x) does not occur in $\mathcal{S}_0$' is in R then we call the variable u (x) *eigenvariable*.

We require that

1. $\mathcal{L}(\mathcal{A}_{i,j}) \subset^* \mathcal{L}(\mathcal{A}_0)$ for all $i = 1 \ldots k$ and $j = 1 \ldots n_i$ (where the symbol $\mathcal{L}(\mathcal{A})$ denotes the set of predicate constants and predicate variables occurring in $\mathcal{A}$, and the symbol $\mathcal{L} \subset^* \mathcal{L}'$ denotes $\mathcal{L} \subset \mathcal{L}' \cup \{\bot, \top, \ldots\}$), and
2. if a formula appears as a side formula in a premise of a rule, then it should appear in the consequent of the rule (and in the same side of the sequent arrow). Any side formula in $\mathcal{S}_0$ should also appear as a side formula in some premise; there is no restriction on the number of side formulas and on their logical form.

From condition 1 it follows that $\mathcal{L}(\mathcal{A}_0) \supset^* \bigcup_{i=1\ldots n, j=1\ldots n_i} \mathcal{L}(\mathcal{A}_{i,j})$. It is worthwhile to stress here that we do not demand that $\mathcal{A}_0$ contain the $\mathcal{A}_i$'s as proper subformulas. We only require $\mathcal{A}_0$ to satisfy condition 1 of the definition of schematic rule (i.e. the predicate symbols in $\mathcal{A}_i$ should appear in $\mathcal{A}_0$). We will refer to condition 1 as *weak embeddability* condition because it is essentially a weaker form of the well-known subformula property.

Rules admitted by the definition are for instance those rules where the auxiliary formulas appear (maybe more than once) as subformulas of A_0. Since $\bot, \top$ are considered as special symbols in the language, we can express moreover Schutte's rule of substitution as a schematic rule. Note that the script letters $\mathcal{A}_0$ and $\mathcal{A}_{i,j}$ we use might denote formulas of any logical form as well as formulas with a prefixed logical form. Examples of schematic rules are

$$\frac{\Gamma \to \Delta, (P \wedge Q) \vee R}{\Gamma \to \Delta, (P \vee R) \wedge (Q \vee R)} \qquad \frac{\Gamma \to \Delta, P \quad \Gamma \to \Delta, Q}{\Gamma \to \Delta, (P \vee R) \wedge (Q \vee R)}$$

In the example on the left the auxiliary formula of the rule is expected to have a certain logical form while in the example on the right the form is arbitrary since any substitution for P and Q is allowed.

Condition 2 expresses minimal requirements for the logical links existing between side formulas lying in the upper and lower sequents. It allows one to distinguish between the rules

$$\frac{\Gamma \to \Delta, P \quad \Theta \to \Lambda, Q}{\Gamma, \Theta \to \Delta, \Lambda, P \wedge Q} \qquad \frac{\Gamma \to \Delta, P \quad \Gamma \to \Delta, Q}{\Gamma \to \Delta, P \wedge Q}$$

where in the left rule there is an intended logical link between all formulas in $\Gamma, \Theta, \Delta, \Lambda$ in the upper sequents and the corresponding copy in the lower sequent; in the right rule, each occurrence in Γ, Δ in the lower sequent is linked to the pair of corresponding occurrences in the upper sequents.

The notion of *step of inference* derived from a schematic rule is defined in the obvious way, similar to the notion of derived axiom. As for schematic axioms we intend collections of side formulas in a schematic rule to be *arbitrary*. Therefore one can derive from the same schematic rule steps of inference which are different exactly because of different side formulas.

A *schematic system* is defined to be a finite set of schematic axioms and schematic rules. The vast majority of systems studied in the literature fall under the notion of schematic system in the sense defined above.

A *proof* in a schematic system $\mathcal{Z}$ is a tree of sequents of *regular* formulas. Each sequent must be derived either by a schematic axiom (in this case the sequent is labelling a leaf of the tree) or by one of the schematic rules of inference in $\mathcal{Z}$ (the sequent is a label for an internal node of the tree). Every occurrence of a sequent in a proof other than the end-sequent is used exactly once as a premise of an inference. The end-sequent of the proof is labelling the root of the tree.

Remark. From our analysis it will turn out that the construction of an interpolant for a sequent is inherently linked to the *logical* rules used in the proof of such a sequent. For this reason without loss of generality we assume to work with schematic systems with built-in permutation and weakening rules. For, we assumed sequents to have the form $\Gamma \rightarrow \Delta$ where Γ, Δ are *collections* of formulas (this assumption avoids explicit permutations), and axioms to admit side formulas both on the right and left hand side of the sequent arrow (this assumption avoids explicit weakenings).

Proposition 1. (Addition of weak occurrences) *If $\Pi : \Gamma \rightarrow \Delta$ is a proof of k lines in a schematic system $\mathcal{Z}$ and Θ, Λ are regular formulas then there is a proof $\Pi' : \Gamma, \Theta \rightarrow \Delta, \Lambda$ of k lines in $\mathcal{Z}$.*

Proof. By induction on the height of a proof. Add suitably Θ, Λ as side occurrences in the axioms of Π.

2.2 Regular Schematic Systems

The Craig Interpolation Theorem will be proved for schematic systems containing a kernel of schematic rules closed under a condition of *regularity*. This condition is natural, and indeed most of the well-known logical systems are exactly defined by a set of rules closed under regularity (i.e. the kernel is the system itself). Given a proof for a certain sequent, an interpolant will be built from distinguished occurrences in the axioms of the proof out of regular rules only. Since the kernel of rules closed under regularity may be *properly* included in the set of rules of the system, the exclusive use of regular rules for building the interpolant implies that the usual symmetry between the logical form of the interpolant and the logical connectives introduced in the proof need not hold when we look at regular schematic systems. It is indeed just a coincidence that the phenomenon holds for specific cases.

A pair of schematic rules is said to be *k-regular* (for $k > 1$) if one has arity k, the other has arity 1, their form is

$$\frac{\mathcal{S}_1[\mathcal{A}_1], \ldots, \mathcal{S}_k[\mathcal{A}_k]}{\mathcal{S}_0[\mathcal{A}_0]} \qquad\qquad \frac{\mathcal{S}[\mathcal{A}'_1, \ldots, \mathcal{A}'_k]}{\mathcal{S}[\mathcal{A}_0]}$$

and the following conditions are satisfied

1. the $\mathcal{A}_i$'s and $\mathcal{A}'_i$'s are atomic formulas formed from predicate variables, variables and metavariables only (for $i = 1 \ldots k$), and
2. $\mathcal{A}_i$ is a variant of $\mathcal{A}'_i$ for all $i = 1 \ldots k$ (i.e. $\mathcal{A}_i$ can be obtained from $\mathcal{A}'_i$ by replacing some of the variables and metavariables in $\mathcal{A}_i$ with different ones), and
3. $\mathcal{L}(\mathcal{A}_0) \subset^* \bigcup_{i=1\ldots k} \mathcal{L}(\mathcal{A}_i)$ (note that $\mathcal{L}(\mathcal{A}_i) = \mathcal{L}(\mathcal{A}'_i)$ by condition 2), and
4. the formula $\mathcal{A}_i$ lies on the right (left) hand side of $\mathcal{S}_i$ whenever $\mathcal{A}'_i$ lies on the left (right) hand side of $\mathcal{S}$ (for $i = 1 \ldots k$); the formula $\mathcal{A}_0$ lies on the right (left) hand side of $\mathcal{S}_0$ whenever $\mathcal{A}_0$ lies on the left (right) hand side of $\mathcal{S}$, and
5. if $\mathcal{A}_0$ does not contain quantifiers then there are no restrictions on the pair of rules; otherwise (i.e. $\mathcal{A}_0$ contains quantifiers) restrictions on eigenvariables (i.e. variables only occurring in $\mathcal{A}_1, \ldots, \mathcal{A}_k$ and $\mathcal{A}'_1, \ldots, \mathcal{A}'_k$ but not occurring in the side formulas of $\mathcal{S}_0, \mathcal{S}$ respectively) are admitted, and each $\mathcal{A}_i$ (for $i = 1 \ldots k$) is obtained from $\mathcal{A}'_i$ precisely by changing some of the variables and metavariables into eigenvariables, and eigenvariables into variables and metavariables.

The k-ary rule is called the *dual* of the unary rule, and vice versa. Rules which belong to a regular pair will be called *regular* rules. It is important to notice that auxiliary formulas of regular rules never have a specific logical form because of condition 1. Condition 3 together with condition 1 in the definition of schematic rule implies that for regular rules $\mathcal{L}(\mathcal{A}_0) =^* \bigcup_{i=1\ldots k} \mathcal{L}(\mathcal{A}_i)$.

A rule is *$\mathcal{Z}$-definable* if either it is a regular rule which belongs to a schematic system $\mathcal{Z}$ or it is obtained by successive applications of regular rules in $\mathcal{Z}$. For instance we say that the rule

$$\frac{\Gamma \to \Delta, P \quad \Gamma \to \Delta, Q}{\neg(P \wedge Q), \Gamma \to \Delta}$$

is LK-definable because obtained by a $\wedge$:*right* rule followed by a $\neg$:*left* in LK.

Let $\mathcal{Z}$ be a schematic system consisting of rules of arity $k_1, \ldots, k_n$. The system $\mathcal{Z}$ is called *regular* if for all $k \in \{k_1, \ldots, k_n\}$ and all choices of $i_0, i_1, \ldots, i_k$ (with $i_j \in \{1, 2\}$), there is a pair of $\mathcal{Z}$-definable k-regular rules such that the k-ary rule of the pair has auxiliary formulas $\mathcal{A}_j^{i_j}$ defined out of predicate variables of arity

0 only. If a system admits restrictions on eigenvariables for some of its rules, we say that it is a *regular* system whenever the above condition is satisfied and moreover for all $n > 0$ it contains pairs of $\mathcal{Z}$-definable unary regular rules with auxiliary formulas $\mathcal{A}_1, \mathcal{A}'_1$ defined out of predicate variables of arity n, and main formula $\mathcal{A}_0$ where none of the variables and metavariables occurring in $\mathcal{A}_1, \mathcal{A}'_1$ appear free.

As emphasised at the beginning of this section, notice that a system closed under regular rules might also contain rules that are not regular.

3 Logical graphs

To trace the logical relations between formula occurrences in a proof Π (i.e. a tree of sequents of regular formulas as defined in section 2) we will be using the notion of *logical edge* introduced in [2] and [7]. We will let positive and negative atomic occurrences in a sequent be logically linked as a rule requires (this is a crucial difference with the definition of logical edges allowed in [2] and [7] where positive occurrences are required to be linked to positive occurrences, and negative to negative). By doing this, we respect those logical conditions on the structure of the rules which are necessary to ensure the construction of an interpolant. To keep track of these logical conditions, even though at the end we will consider only graphs based on logical edges between *atomics*, we will define logical graphs also on those logical links that are induced by schematic rules and schematic axioms (notice that atomic formulas in schematic axioms and schematic rules may lead to non-atomic formulas after substitution).

An *s-formula* is an *occurrence* of a subformula of a formula (here, '*s*-' stands for '*semi-*' or '*sub-*'). It has to be emphasized that an *s*-formula is an *occurrence* of a subformula (in the proof) as opposed to the subformula itself which may occur many times (in the proof). A formula $\mathcal{A}$ is a *variant* of $\mathcal{B}$ if $\mathcal{A}$ can be obtained from $\mathcal{B}$ by replacing some of the terms and metaterms in $\mathcal{B}$ with different ones. If δ is a substitution map and $\delta(\mathcal{A}), \delta(\mathcal{B})$ are defined, it is easy to check that $\delta(\mathcal{A})$ is a variant of $\delta(\mathcal{B})$ whenever $\mathcal{A}$ is a variant of $\mathcal{B}$.

Two *s*-formulas in a proof will be connected by an edge only if they are variants of each other; any two *s*-formulas connected by an edge will occur respectively in an upper and lower sequent of some rule, or will both be in an axiom on opposite sides of the sequent arrow. More precisely, they will be defined as follows.

First, an axiom in Π is derived with some substitution δ from a schematic axiom containing two distinguished formulas $\mathcal{A}_1, \mathcal{A}_2$. By definition of schematic axiom either all atomic *s*-formulas $\mathcal{B}$ (which are not special constants) in $\mathcal{A}_1$ have variants $\mathcal{B}_1, \ldots \mathcal{B}_n$ (for some $n \geq 1$) in $\mathcal{A}_2$ (i.e. $\mathcal{L}(\mathcal{A}_1) \subset^* \mathcal{L}(\mathcal{A}_2)$) or the other way around (i.e. $\mathcal{L}(\mathcal{A}_2) \subset^* \mathcal{L}(\mathcal{A}_1)$). Define an edge between each formula $\delta(\mathcal{B})$ and some of its variants $\delta(\mathcal{B}_i)$ (maybe all but at least one). If an axiom is derived from a schematic axiom containing only one distinguished occurrence, there will be no edges defined.

Second, in any schematic rule all atomic s-formulas $\mathcal{B}$ in auxiliary formulas must have variants $\mathcal{B}_1, \ldots \mathcal{B}_n$ (for some $n \geq 1$) in the main formula, except for special constants. Suppose a step of inference in Π is derived with a substitution δ from some schematic rule. We define an edge between each s-formula $\delta(\mathcal{B})$ and some of its variants $\delta(\mathcal{B}_i)$ (maybe all but at least one; we do not ask that *all* possible variants in the main formula be linked to $\delta(\mathcal{B})$).

Third, for any side formula $\mathcal{A}$ in the upper sequent(s) of a schematic rule, by condition 2 (in the definition of schematic rule) there is a side formula $\mathcal{A}$ in the lower sequent. Suppose a rule is derived with some substitution δ from a schematic rule. We define an edge between each side formula $\delta(\mathcal{A})$ occurring in the *upper* sequent(s) and the side formula $\delta(\mathcal{A})$ occurring in the *lower* sequent.

Fourth, suppose there is an edge from an s-formula A_1 to A_2 in Π (A_1, A_2 are regular formulas because they occur in Π) and suppose B_1 is an atomic subformula of A_1. Since A_1 and A_2 are variants, there is a subformula B_2 of A_2 which correspond to the subformula B_1 of A_1; the s-formulas B_1 and B_2 are of course variants. Define an edge between B_1 and B_2.

The *logical graph* of a proof is a graph obtained by tracing the edges between the atomic s-formulas in the proof as described above (and simply forgetting afterwards all those edges connecting formulas which are not atomic).

Note that more than one s-formula in the auxiliary formula(s) of a rule might have an edge to the same variant in the main formula and more than one s-formula in the main formula might have edges to the same s-formula in the antecedent(s). Therefore a connected subgraph of the logical graph for a schematic proof is not necessarily a tree.

It should be stressed that no rules having logical edges defined between formulas in upper sequents (e.g. the cut rule), belong to schematic systems we consider.

A path in a proof is defined as a sequence of consecutive edges in the logical graph of the proof such that there are no two edges of the path crossing the same axiom or the same rule of inference. A path is *full* when it is maximal in the logical graph of a proof where it lies. A path is *direct* if it does not cross any axiom.

Proposition 2. *Let Π be a proof of $A \to B$ and C be an atomic s-formula (which is not a special predicate constant) in $A \to B$. All full paths from C in the logical graph of Π are either direct or they go back to a variant in $A \to B$.*

Proof. A full path from C either crosses an axiom (in which case it goes back to the end-sequent) or it does not (in which case it is a direct path).

Proposition 3. *Let Π be a proof of $A \to B$ and C be a formula occurrence in Π. Then either all direct paths from atomic s-formulas (which are not special predicate constants) in C down to the end-sequent go to A or they all go to B.*

Proof. The main point is that while an atomic occurrence in the upper sequent of a rule may be linked to more than one atomic occurrence in the lower sequent, all its linked occurrences in the lower sequent lie in the same formula.

If all direct paths from atomic s-formulas of C in Π go to A (B) we say that the formula *goes* to A (B).

Notice that Proposition 3 does *not* imply the well known *subformula property* ([14]). Clearly, any schematic system that satisfies the subformula property is a schematic system satisfying Proposition 3. As we will see in the conclusion, our proof of the Craig Interpolation Theorem does not need the subformula property to hold.

4 The Statement and the Proof

We are now ready to formalize the statement of Craig's Theorem for regular schematic systems.

Theorem 4. (Craig's Interpolation Theorem for Regular Schematic Systems) *Let $\mathcal{Z}$ be a regular schematic system. Let A and B be two regular formulas such that the sequent $A \to B$ is provable in $\mathcal{Z}$, and that A and B have at least one predicate constant in common. Then there exists a formula C (called the interpolant of $A \to B$) containing only those predicate constants that occur in both A and B and possibly special predicate constants, such that $A \to C$ and $C \to B$ are provable in $\mathcal{Z}$.* [3]

The interpolant will be built by steps on the complexity of the subproofs using some easy considerations on the logical flows of formulas in the proof. The method brings to light the logical relations between the s-formulas occurring in the interpolant and the ones occurring in A and B. Regular rules will be the only rules used to build the interpolant C at each stage of the construction.

The technique was used to prove the theorem for LK and LJ in [3].

4.1 Proof of the Interpolation Theorem

Let Π_* be a proof of $A \to B$. The interpolant C_* for $A \to B$ associated to Π_* together with the proofs $\Pi_*^A : A \to C_*$ and $\Pi_*^B : C_* \to B$ will be built in stages from the subproofs of Π_*. The idea consists of 'extracting' from the proof Π_*, two proofs associated to A and B respectively. As a consequence, only part of the original proof Π_* that 'contributes' to the construction of the formula A (B respectively) will appear in the proof Π_*^A (Π_*^B respectively).

Proof. (Theorem 4) We will build a formula C_* and proofs $\Pi_*^A : A \to C_*, \Pi_*^B : C_* \to B$ in $\mathcal{Z}$. To indicate that the language of C_* (i.e. the set of symbols occurring in C_*) is contained in the intersection of the language of A with the language of B extended with $\top, \bot$, we use the symbol $\mathcal{L}(C_*) \subset^* \mathcal{L}(A) \cap \mathcal{L}(B)$.

The proof is by induction on the height of the subproofs Π of Π_*. For each subproof Π of Π_* with end-sequent $\Gamma \to \Delta$ we show that

[3] A stronger statement where we require that variables, function symbols and constants in C must appear in both A and B, can be given. But in this case some extra conditions on terms appearing in rules are needed.

1. there exist subcollections Γ^A, Δ^A and Γ^B, Δ^B of Γ, Δ such that
   a. each formula occurrence in Γ^A, Δ^A (Γ^B, Δ^B, respectively) goes to A (B, respectively), and
   b. Γ is the collection obtained by combining Γ^A, Γ^B; similarly Δ is defined by combining Δ^A, Δ^B;
2. there exists a formula C_Π such that
   a. $\mathcal{L}(C_\Pi) \subset^* \mathcal{L}(A) \cap \mathcal{L}(B)$, and
   b. $\Gamma^A \rightarrow \Delta^A, C_\Pi$ and $C_\Pi, \Gamma^B \rightarrow \Delta^B$ are provable sequents.

Notice that if Π is Π_* then Γ^A is A, Δ^A is $\emptyset$, Γ^B is $\emptyset$, Δ^B is B and $A \rightarrow C_*$, $C_* \rightarrow B$ turn out to be provable sequents (following the notation, C_* is precisely C_{Π_*}).

Let us start first by considering subproofs Π of Π_*, of height 1 of the form $D, \Theta \rightarrow \Lambda, D'$. There are four cases we should consider:

1. the occurrence D goes to A in Π_*, while D' goes to B; if $\mathcal{L}(D) \subset^* \mathcal{L}(D')$ then let C_Π be D, the proof Π^A be $D, \Theta^A \rightarrow \Lambda^A, D$, the proof Π^B be $D, \Theta^B \rightarrow \Lambda^B, D'$ and the subcollections $\Gamma^A, \Delta^A, \Gamma^B, \Delta^B$ be the pairs D, Θ^A, Λ^A, D, D, Θ^B, Λ^B, D respectively (in the following we will not mention anymore how subcollections are defined; they will be clear from the context). If $\mathcal{L}(D') \subset^* \mathcal{L}(D)$ then let C_Π be D', the proof Π^A be $D, \Theta^A \rightarrow \Lambda^A, D'$ and Π^B be $D', \Theta^B \rightarrow \Lambda^B, D'$.
2. the occurrence D goes to B, while D' goes to A; if $\mathcal{L}(D) \subset^* \mathcal{L}(D')$ then let C_Π be D, take Π^A to be $D, \Theta^A \rightarrow \Lambda^A, D'$, and Π^B to be $D, \Theta^B \rightarrow \Lambda^B, D$; if $\mathcal{L}(D') \subset^* \mathcal{L}(D)$ then let C_Π be D', take Π^A to be $D', \Theta^A \rightarrow \Lambda^A, D'$, and and Π^B to be $D, \Theta^B \rightarrow \Lambda^B, D'$.
3. both occurrences D, D' go to A; let C_Π be $\bot$, take Π^A to be $D, \Theta^A \rightarrow \Lambda^A, D', \bot$ and Π^B to be $\bot, \Theta^B \rightarrow \Lambda^B$.
4. both occurrences D, D' go to B; let C_Π be $\top$, take Π^A to be $\Theta^A \rightarrow \Lambda^A, \top$ and Π^B to be $\top, D, \Theta^B \rightarrow \Lambda^B, D'$.

In case a schematic axiom contains only one distinguished occurrence we proceed in a similar spirit and say that if the schematic axiom has the form $\Theta \rightarrow \Lambda, D'$ with D' going to B (A) then Π^A is $\Theta^A \rightarrow \Lambda^A, \top$, the proof Π^B is $\top, \Theta^B \rightarrow \Lambda^B, D'$ and C_Π is $\top$. Similarly if the schematic axiom has the form $D, \Theta \rightarrow \Lambda$ with D going to A (B).

For subproofs Π of height greater than 1 we will examine the last rule of inference R. Suppose R is a k-ary (for $k \geq 1$) rule applied to subproofs $\Pi_1 \dots \Pi_k$

$$\frac{\begin{array}{ccc}\Pi_1 & & \Pi_k \\ \mathcal{S}_1[A_{1,1}, \dots, A_{1,n_1}] & \dots & \mathcal{S}_k[A_{k,1}, \dots, A_{k,n_k}]\end{array}}{\mathcal{S}_0[A_0]}$$

with eigenvariables $b_1 \dots b_r$ (with $r \geq 0$) of Π. By induction hypothesis there are k pairs of proofs Π_i^A and Π_i^B for $i = 1 \dots k$ and k formulas $C_{\Pi_1}, \dots, C_{\Pi_k}$ such that $\mathcal{L}(C_{\Pi_i}) \subset^* \mathcal{L}(A) \cap \mathcal{L}(B)$ for $i = 1 \dots k$. We will use this information to build the formula C_Π and, $\Pi^A : \mathcal{S}^A[C_\Pi]$ and $\Pi^B : \mathcal{S}^B[C_\Pi]$. There are two cases:

1. the main formula A_0 goes to A. Let C_Π be C_0 and define Π^A to be

$$\cfrac{\cfrac{\begin{array}{ccc}\Pi^A_{1,*} & & \Pi^A_{k,*} \\ \mathcal{S}^A_1[A_{1,1},\ldots,A_{1,n_1},C'_{\Pi_1}] & \ldots & \mathcal{S}^A_k[A_{k,1},\ldots,A_{k,n_k},C'_{\Pi_k}]\end{array}}{\mathcal{S}^A[A_0,C'_{\Pi_1},\ldots,C'_{\Pi_k}]}}{\mathcal{S}^A[A_0,C_0]}$$

where $\Pi^A_{1,*}\ldots\Pi^A_{k,*}$ are subproofs obtained from $\Pi^A_1\ldots\Pi^A_k$ by applying a regular unary rule to $C_{\Pi_1}\ldots C_{\Pi_k}$ to eliminate all eigenvariables, for all $i = 1\ldots k$. If C_{Π_i} does not contain eigenvariables, $\Pi^A_{i,*}$ is defined to be Π^A_i. Notice that the regular unary rule applied to C_{Π_i} (if needed) exists (to be precise, it is $\mathcal{Z}$-definable) because the system is assumed to be regular.
We assume it is clear from the context what are the formula occurrences which are auxiliary and/or main formulas of each rule application. In the figure we explicitly displayed all those formula occurrences relevant to the construction and not just the ones relevant to a single rule.
Suppose $\Pi^A_{i,*}$ is different than Π^A_i for some $i = 1\ldots k$. Let Π^B be

$$\cfrac{\begin{array}{c}\Pi^B_{i,*} \\ \mathcal{S}^B_{i,*}[C'_{\Pi_1},C'_{\Pi_2}\ldots C'_{\Pi_k}]\end{array}}{\mathcal{S}^B[C_0]}$$

where $\Pi^B_{i,*}$ is obtained by applying Proposition 1 to Π^B_i so to add to the sequent $\mathcal{S}^B_i$ as weak occurrences, the formulas C'_{Π_j}'s and all formulas in the end-sequents $\mathcal{S}^B_j$ of Π^B_j except the main formulas C_{Π_j}'s (for $j = 1\ldots i-1, i+1\ldots k$), and by applying afterwards a unary regular rule (dual to the one used in Π^A_i) to C_{Π_i} to eliminate all eigenvariables occurring in it. Take C_Π to be C_0. Clearly, $\mathcal{L}(C_0) \subset^* \mathcal{L}(A) \cap \mathcal{L}(B)$.
If no eigenvariables appear in C_{Π_i} of Π^A_i, Π^B_i for all $i = 1\ldots k$, notice that C'_Π is C_Π. The construction of Π^B is the obvious simplification of the one just discussed.
2. the main formula A_0 goes to B. Suppose C_{Π_i} contains an eigenvariable of Π. Let C_Π be C_0 and define Π^A to be

$$\cfrac{\begin{array}{c}\Pi^A_{i,*} \\ \mathcal{S}^A_{i,*}[C'_{\Pi_1},C'_{\Pi_2}\ldots C'_{\Pi_k}]\end{array}}{\mathcal{S}^A[C_0]}$$

where $\Pi^A_{i,*}$ is obtained by applying in a suitable way Proposition 1 to Π^A_i and a unary regular rule to eliminate all eigenvariables from C_{Π_i}. Let Π^B be

$$\cfrac{\cfrac{\begin{array}{ccc}\Pi^B_{1,*} & & \Pi^B_{k,*} \\ \mathcal{S}^B_1[A_{1,1},\ldots,A_{1,n_1},C'_{\Pi_1}] & \ldots & \mathcal{S}^B_k[A_{k,1},\ldots,A_{k,n_k},C'_{\Pi_k}]\end{array}}{\mathcal{S}^B[A_0,C'_{\Pi_1},\ldots,C'_{\Pi_k}]}}{\mathcal{S}^B[A_0,C_0]}$$

where $\Pi^B_{1,*} \ldots \Pi^B_{k,*}$ are defined similarly to case 1. Notice that the regular rule applied to Π^B_i should be the dual of the one applied to Π^A_i.

This concludes the proof.

Remark. Our definition of regularity requires regular pairs to satisfy a *minimal* set of conditions sufficient to prove Theorem 4. Because of this generality, the number of regular pairs required to belong to such a set of rules might be quite large. Theorem 4 can be carried out for a restricted number of regular pairs whenever constraints on the rules used in a proof are given. For instance if we can prove $A \rightarrow B$ with rules whose auxiliary formulas and main formulas both belong either to the right or to the left hand side of a sequent, then we can ask the formulas C_Π in Π^A_1 (Π^B_1) to always belong to the right (left) hand side of a sequent, respectively. Therefore we would only need to consider regular rules whose arguments and values lie both either in the left or in the right hand side of the sequents.

In the sequent calculus for classical logic LK (for a reference see [14]) all pairs of rules introducing a connective to the right and to left hand side of the sequent arrow, form a regular pair. Because of Theorem 4 this means that LK enjoys the interpolation property. For the same reason also the LK-fragments $[\neg]$, $[\wedge, \vee, \neg, \supset]$, $[\wedge]$, $[\vee]$, $[\wedge, \vee]$ satisfy the property (notice that for the last three fragments we should apply the discussion in our latter remark). For the fragments $[\wedge, \neg]$ and $[\vee, \neg]$ we should observe that regularity follows from $\mathcal{Z}$-definability, for $\mathcal{Z}$ being $[\wedge, \neg]$ and $[\vee, \neg]$ respectively. Last, let us remark that Schutte system for the treatment of formula substitutions enjoys the property as well as the fragments of linear logic LL of type I (i.e. elementary fragments without linear negation and without multiplicative implication) and II (i.e. elementary fragments without negation or without multiplicative implication together with 0 (falsum)). This was proved also in [12]. This list of calculi is of course not exhaustive.

Remark. There is no assumption on the number of formulas occurring on the right and left hand side of a schematic sequent. Notice that the intuitionistic calculus LJ is formalized by assuming that sequents contain *at most* one formula occurrence on the right of a sequent. Our result can be easily modified for regular systems satisfying such requirement. The central idea consists in building the interpolant by suitably switching its position in the sequent during the construction. To do this, one needs some extra conditions to the notion of regular schematic system which we will not discuss here.

5 Conclusions

Let us briefly point out some directions for further inquiry.

As a result of our analysis we now have a number of conditions sufficient for a logical system to enjoy Interpolation. Hopefully these conditions will provide new insights for the automatic analysis of properties of logical systems (work in this direction was done in [1] for multi-valued logics).

Theorem 4 shows that a logical system need not satisfy the *subformula property* for its rules in order to enjoy Interpolation, but simply some weak form of *embedding* of the antecedents of a rule into its consequent. We want to remark here that our definition of embedding can be weakened so to interpret logical systems formalizing proofs with analytic cuts (i.e. cuts on formulas appearing as subformulas of the endsequent of the proof) which are known to enjoy Interpolation (the reader can refer to [4] for results in this direction).

The notion of schematic systems we use together with notions like weak embeddability and *Z-definability* of rules of inference seem useful to characterize *similarity* of proofs, i.e. similar proofs of similar conclusions. There are different approaches to the question of similarity already in the literature (see [8], [11], [13]) and a definite mathematical concept capturing the intuition of *similarity* has not been obtained yet.

References

1. M. Baaz, C.G. Fermüller and R. Zach. Systematic construction of natural deduction systems for many-valued logics. *Proc. 23rd International Symposium on Multiple-valued Logics*, Sacramento CA, IEEE Press , 208–213, 1993.
2. S. Buss. The undecidability of k-provability. *Annals of Pure and Applied Logic*, 53:72–102, 1991.
3. A. Carbone. *On Logical Flow Graphs.* Ph.D. Dissertation, Department of Mathematics of the Graduate School of The City University of New York, May 1993.
4. A. Carbone. Interpolants, Cut Elimination and Flow Graphs. To appear in *Annals of Pure and Applied Logic*, 1995.
5. W. Craig. Linear Reasoning. A new form of the Herbrand-Gentzen Theorem. In *Journal of Symbolic Logic*, 22:250–268, 1957.
6. W. Farmer. A unification-theoretic method for investigating the k-provability problem. In *Annals of Pure and Applied Logic*, 51:173–214, 1991.
7. J.Y. Girard. Linear Logic. In *Theoretical Computer Science,* 50:1–102, 1987.
8. G. Kreisel. A survey of proof theory II. In J.E. Fenstad, editor, *Proceedings Second Scandinavian Logic Symposium*, pages 109–170. 1971.
9. J. Krajíček. On the number of steps in proofs. In *Annals of Pure and Applied Logic*, 41:153–178, 1989.
10. R. Parikh. Some results on the length of proofs. In *Transaction of the American Mathematical Society*, 177:29–36, 1973.
11. D. Prawitz. Ideas and results in proof theory. In J.E. Fenstad, editor, *Proceedings Second Scandinavian Logic Symposium*, pages 235–307. 1971.
12. D. Roorda. *Resource Logics: Proof-theoretical Investigations.* Ph.D. Dissertation, Department of Mathematics and Computer Science of the University of Amsterdam, September 1991.
13. M.E. Szabó. *Algebra of Proofs.* North-Holland, Amsterdam, 1977.
14. G. Takeuti. *Proof Theory.* Studies in Logic 81. North Holland, Amsterdam, 2nd Edition, 1987.

The Role of Monotonicity in Descriptive Complexity Theory

Iain A. Stewart

University of Wales Swansea, Swansea SA2 8PP, U.K.

Abstract. It is well-known that monotonicity, in a variety of guises, plays an essential role in descriptive complexity theory and finite model theory. In this paper, we highlight existing well-known roles and also explain two other not so well-known roles of monotonicity. The first arises due to an anomaly in logically characterizing NP using operators corresponding to traditional NP-complete problems; and the second arises due to the consideration of the logical characterization of complexity classes defined with respect to "monotone" models of computation.

1 Introduction

It is well-known that monotonicity, in a variety of guises, plays an essential role in descriptive complexity theory and finite model theory. In this paper, we highlight existing well-known roles, namely

- in the formation of formulae of least-fixed point logic;
- in the playing of extended Ehrenfeucht-Fraïssé games;
- in database query languages like Datalog,

and also explain two other not so well-known roles of monotonicity. The first of these roles arises due to an anomaly in logically characterizing NP using uniform sequences of Lindström quantifiers (or operators) corresponding to traditional NP-complete problems; and the second arises due to the consideration of the logical characterization of complexity classes defined with respect to "monotone" models of computation.

2 Basic Definitions

As full definitions of most concepts can be found, for example, in [18], we only present outlines here. Throughout the paper a signature τ is a tuple of relation and constant symbols (although we sometimes insist that our signatures be purely relational). All structures S over τ are finite with universe $|S| = \{0, 1, \ldots, n-1\}$, for some positive integer n. The size of some structure S is also denoted by $|S|$. We denote the set of all structures over τ by STRUCT(τ) and we assume that all structures are of size at least 2. A *problem of arity t* (≥ 0) over τ is a subset of $\mathrm{STRUCT}_t(\tau) = \{(S, \mathbf{u}) : S \in \mathrm{STRUCT}(\tau), \mathbf{u} \in |S|^t\}$

that is closed under isomorphism. If Ω is some problem then $\tau(\Omega)$ denotes its signature.

The language of the first-order logic $FO_s(\tau)$ is as expected except that it has a built-in successor relation s and two constant symbols 0 and max that are always interpreted as 0 and $n-1$, respectively, in any universe of size n (if we disallow the built-in successor relation s in first-order logic then we denote the resulting logic by FO). Any formula ϕ of $FO_s(\tau)$ with free variables those of the t-tuple $\mathbf{x}$ is interpreted in the set $STRUCT_t(\tau)$. (We usually write $s(x,y)$ as $y = x+1$ and $\neg(x = y)$ as $x \neq y$.) Also, $FO_s = \cup\{FO_s(\tau) : \tau \text{ some signature}\}$. A formula ϕ *describes* the problem

$$\{(S,\mathbf{u}) : (S,\mathbf{u}) \in STRUCT_t(\tau), (S,\mathbf{u}) \models \phi(\mathbf{x})\}$$

of arity t.

Let τ_2 be the signature consisting of the binary relation symbol E: so, we may clearly consider structures over τ_2 as digraphs or graphs. Consider the problem HP of arity 2 defined as follows:

$$HP = \{(S,u,v) \in STRUCT_2(\tau_2) : \text{there is a Hamiltonian path in the digraph } S \text{ from vertex } u \text{ to vertex } v\}.$$

Suppose that $\phi(\mathbf{x},\mathbf{y})$ is a formula of FO_s where $\mathbf{x}$ and $\mathbf{y}$ are k-tuples of distinct variables. Then the digraph denoted by $HP[\lambda\mathbf{x},\mathbf{y}\phi(\mathbf{x},\mathbf{y})]$ has its vertices indexed by k-tuples (over the domain of the structure in which it is interpreted) and there is an edge from vertex $\mathbf{u}$ to vertex $\mathbf{v}$ if and only if there is a Hamiltonian path from $\mathbf{u}$ to $\mathbf{v}$ in the digraph described by the relation $\phi(\mathbf{x},\mathbf{y})$. Consequently, the formula $HP[\lambda\mathbf{x},\mathbf{y}\phi(\mathbf{x},\mathbf{y})](\mathbf{0},\mathbf{max})$ describes all those structures (over the signature in question) for which there is a Hamiltonian path from vertex $\mathbf{0}$ to vertex $\mathbf{max}$ in the digraph described by $\phi(\mathbf{x},\mathbf{y})$ ($\mathbf{0}$ (resp. $\mathbf{max}$) is the constant symbol 0 (resp. max) repeated k times). Clearly, we can re-apply the operator HP to formulae already containing applications of HP. If we allow an unlimited number of nested applications of the operator HP then we denote the resulting logic by $(\pm HP)^*[FO_s]$ (this is the logic (FO+HP) of [17]). We write $(\pm HP)^k[FO_s]$ (resp. $HP^*[FO_s]$) to denote the sub-logic of $(\pm HP)^*[FO_s]$ where all formulae have at most k nested applications of the operator HP (resp. where no operator appears within the scope of a negation sign; that is, it appears *positively*): the sub-logic $HP^k[FO_s]$ of $HP^*[FO_s]$ is defined similarly. Needless to say, first-order logic can be extended by other operators.

Let $\tau' = \langle R_1, R_2, \ldots, R_r, C_1, C_2, \ldots, C_c\rangle$ be some signature where each R_i is a relation symbol of arity a_i and each C_j is a constant symbol, and let $L(\tau)$ be some logic over some signature τ. Then the set of formulae $\Sigma = \{\phi_i(\mathbf{x}_i), \psi_j(\mathbf{y})_j : i = 1,2,\ldots,r; j = 1,2,\ldots,c\} \subseteq L(\tau)$, where:

- each formula ϕ_i is over the qa_i distinct variables $\mathbf{x}_i$, for some fixed positive integer q, and each formula $\psi_j(\mathbf{y}_j)$ is over the q distinct variables $\mathbf{y}_j$;
- for all $S \in STRUCT(\tau)$ and for all $j = 1,2,\ldots,c$, there exists exactly one tuple $\mathbf{u} \in |S|^q$ such that $(S,\mathbf{u}) \models \psi_j(\mathbf{y}_j)$,

is called τ'*-descriptive* (note that if τ_0, say, is a different signature to τ' and Σ contains formulae corresponding to the relation symbols of τ_0, then the set of formulae Σ would be called τ_0-descriptive). For each $S \in \text{STRUCT}(\tau)$, the τ'-*translation of* S *with respect to* Σ *(of arity* q*)* is the structure $S' \in \text{STRUCT}(\tau')$ with universe $|S|^q$ defined as follows: for all $i = 1, 2, \ldots, r$ and for any tuples $\{\mathbf{u}_1, \mathbf{u}_2, \ldots, \mathbf{u}_{a_i}\} \subseteq |S'| = |S|^q$:

$$R_i^{S'}(\mathbf{u}_1, \mathbf{u}_2, \ldots, \mathbf{u}_{a_i}) \text{ holds if and only if } (S, (\mathbf{u}_1, \mathbf{u}_2, \ldots, \mathbf{u}_{a_i})) \models \phi_i(\mathbf{x}_i),$$

and for all $j = 1, 2, \ldots, c$:

$$C_j^{S'} = \mathbf{u} \text{ if and only if } (S, \mathbf{u}) \models \psi_j(\mathbf{y}_j).$$

Let Ω and Ω' be problems of arity 0 over the vocabularies τ and τ', respectively. Let Σ be a τ'-descriptive set of formulae from some logic $L(\tau)$ and for each $S \in \text{STRUCT}(\tau)$, let $\sigma(S) \in \text{STRUCT}(\tau')$ denote the τ'-translation of S with respect to Σ. Then Ω' is an L*-translation* of Ω if and only if for each $S \in \text{STRUCT}(\tau)$, $S \in \Omega$ if and only if $\sigma(S) \in \Omega'$. We clearly have the notion of one problem being a *first-order translation* of another.

Define the problems HP(0,max) and 3COL as follows:

$$\begin{aligned} \text{HP(0,max)} &= \{S \in \text{STRUCT}(\tau_2) : \text{there is a Hamiltonian path in the} \\ &\qquad \text{digraph } S \text{ from vertex 0 to vertex } max\}; \\ \text{3COL} &= \{S \in \text{STRUCT}(\tau_2) : \text{the graph } S \text{ can be 3-coloured}\}. \end{aligned}$$

Theorem 1. *(i)* $NP = HP^1[FO_s] = HP^*[FO_s]$ ([17]).

(ii) $NP = 3COL^1[FO_s]$ *but* $3COL^*[FO_s] = L^{\text{NP}}$ ([15, 18, 19, 22]).

(iii) HP(0,max) and 3COL are complete for NP via first-order translations ([15, 17, 22]).

Note that we use, for example, $\text{HP}^1[\text{FO}_s]$ to denote a logic and also the problems describable by the sentences of that logic, and we assume from now on that the problem HP is defined as

$$\{S \in \text{STRUCT}(\tau_2) : \text{there is a Hamiltonian path in the digraph } S \text{ from } 0 \text{ to } max\},$$

and all problems are of arity 0.

3 Some Fundamental Roles of Monotonicity

In this section we describe different well-known notions of monotonicity and highlight how these concepts play a role in finite model theory.

The monotonicity of a formula plays an essential role in the formation of formulae of $(\pm\text{LFP})^*[\text{FO}_s]$, first-order logic augmented with a *least fixed point operator*. Formulae of $(\pm\text{LFP})^*[\text{FO}_s]$ are built as follows:

- any formula of FO_s is a formula of $(\pm\text{LFP})^*[\text{FO}_s]$;

- $(\pm\text{LFP})^*[\text{FO}_s]$ is closed under first-order constructions;
- if $\phi(x_1,\ldots,x_r,R)$ is a formula of $(\pm\text{LFP})^*[\text{FO}_s]$ whose free variables contain those of $\mathbf{x}=(x_1,\ldots,x_r)$; where R is a relation symbol of arity r not in the underlying signature; and where R only appears positively in ϕ (i.e., not within the scope of a negation sign) then the formula

$$\text{LFP}[\lambda\mathbf{x}\phi](\mathbf{y})$$

is a formula of $(\pm\text{LFP})^*[\text{FO}_s]$, where $\mathbf{y}=(y_1,\ldots,y_r)$ are variables or constant symbols (the variables of $\mathbf{x}$ are bound in the new formula and those of $\mathbf{y}$ are free: there may also be other free variables in ϕ and $\text{LFP}[\lambda\mathbf{x}\phi](\mathbf{y})$).

A sentence $\text{LFP}[\lambda\mathbf{x}\phi](\mathbf{y})$, as above, of $(\pm\text{LFP})^*[\text{FO}_s]$ is interpreted in some structure S, over the underlying signature of ϕ, as follows. Let R_0 be the r-ary empty relation over $|S|$ and for all $i\geq 0$, define:

$$R_{i+1}=\{\mathbf{u}\in|S|^r:\phi^S(\mathbf{u},R_i)\text{ holds}\}.$$

Then $R_i\subseteq R_{i+1}$ and so ϕ has a least fixed-point R_f. The structure S satisfies $\text{LFP}[\lambda\mathbf{x}\phi](\mathbf{y})$ if, and only if, the r-tuple over $|S|$ corresponding to $\mathbf{y}$ is in R_f.

The condition, above, that the relation symbol R appears only positively in $\phi(\mathbf{x},R)$ ensures that $\phi(\mathbf{x},R)$ is *monotone*; that is, for any structure S over the underlying signature of ϕ and any pair of r-ary relations R_1 and R_2 on $|S|$, if $R_1\subseteq R_2$ then:

$$\{\mathbf{u}\in|S|^r:\phi^S(\mathbf{u},R_1)\text{ holds}\}\subseteq\{\mathbf{u}\in|S|^r:\phi^S(\mathbf{u},R_2)\text{ holds}\}.$$

Consequently, all problems defined by sentences of $(\pm\text{LFP})^*[\text{FO}_s]$ are solvable in polynomial time. Immerman [10] and Vardi [24] proved that $(\pm\text{LFP})^*[\text{FO}_s]$ actually captures P.

Note that there exist first-order formulae $\phi(\mathbf{x},R)$ that are monotone but not equivalent to any first-order formulae $\phi'(\mathbf{x},R)$ in which R only appears positively. In fact, Ajtai and Gurevich [3] proved that it is undecidable as to whether a first-order formula $\phi(\mathbf{x},R)$ is monotone. If we drop the stipulation that R only appears positively in $\phi(\mathbf{x},R)$ in the definition above, and define the semantics accordingly, then the resulting logic, denoted $(\pm\text{PFP})^*[\text{FO}_s]$, *partial fixed-point logic*, captures PSPACE [24].

An alternative role of monotonicity appears in the extended Ehrenfeucht-Fraïssé games of Kolaitis and Vardi [12] and Grädel [9]. Actually, the games presented below extend those in [9], where the problem Ω is fixed as TC, and differ slightly from those in [12], where the Ω_k moves are defined slightly differently; all signatures are devoid of constant symbols; and the logics involved are extensions of the infinitary logic $\mathcal{L}^\omega_{\infty\omega}$. In fact, the games in [12] were originally defined with respect to families of monotone unary generalized quantifiers and it was claimed (in the final paragraph of Section 3 of [12]) that the relevant proofs work for families of monotone Lindström quantifiers when modified only via notational changes.

Let Ω be some problem over the signature $\tau = \langle R_1, \ldots, R_r, C_1, \ldots, C_c \rangle$, where each R_i is a relation symbol of arity a_i and each C_j is a constant symbol. Our new games, on two structures $\mathcal{A}$ and $\mathcal{B}$ over the same signature, extend the classical Ehrenfeucht-Fraïssé game (see, for example, [12]) as follows. As well as the usual $\exists$- and $\forall$-moves, there are also Ω_k- and $\neg\Omega_k$-moves.

Ω_k-move. The Spoiler selects a τ-structure $\mathcal{S}_\mathcal{A}$ such that the domain of $\mathcal{S}_\mathcal{A}$ is $|\mathcal{A}|^k$; the tuples from $|\mathcal{A}|^k$ determining the constants $C_1^{\mathcal{S}_\mathcal{A}}, \ldots, C_c^{\mathcal{S}_\mathcal{A}}$ consist only of previously pebbled elements of $|\mathcal{A}|$ or constants of $\mathcal{A}$; and $\mathcal{S}_\mathcal{A} \in \Omega$. The Duplicator replies by selecting a τ-structure $\mathcal{S}_\mathcal{B}$ such that the domain of $\mathcal{S}_\mathcal{B}$ is $|\mathcal{B}|^k$; the tuples from $|\mathcal{B}|^k$ determining the constants $C_1^{\mathcal{S}_\mathcal{B}}, \ldots, C_c^{\mathcal{S}_\mathcal{B}}$ of $\mathcal{S}_\mathcal{B}$ consist of the tuples corresponding (in the natural way and componentwise) to those selected by the Spoiler above; and $\mathcal{S}_\mathcal{B} \in \Omega$. Then the Spoiler places, for some $i \in \{1, 2, \ldots, r\}$, ka_i (as yet unused) pebbles on some tuple $\mathbf{u}_i \in |\mathcal{B}|^{ka_i}$ such that $R_i^{\mathcal{S}_\mathcal{B}}(\mathbf{u}_i)$ holds, and the Duplicator replies by placing the corresponding ka_i pebbles on some tuple $\mathbf{v}_i \in |\mathcal{A}|^{ka_i}$ such that $R_i^{\mathcal{S}_\mathcal{A}}(\mathbf{v}_i)$ holds ($\mathbf{u}_i$ and $\mathbf{v}_i$ are considered as a_i-tuples of k-tuples).
$\neg\Omega_k$-move. As for the Ω_k-move but with the roles of $\mathcal{A}$ and $\mathcal{B}$ reversed.

For some string ω over $\{\exists, \forall\} \cup \{\Omega_k, \neg\Omega_k : k \text{ a positive integer}\}$, the *$\Omega(\omega)$-game on $\mathcal{A}$ and $\mathcal{B}$* is played as usual by the Spoiler and the Duplicator, with the winning conditions as for the classical Ehrenfeucht-Fraïssé game, but possibly using the new moves defined above, except with the proviso that if $\omega = \omega_1\omega_2\ldots\omega_t$ then there are t moves in the game and the ith move of the game is a ω_i-move, for all $i = 1, 2, \ldots, t$ (there are always exactly the right number of pebbles so that all t moves can be played).

A problem Ω over the signature τ is *monotone* if for every pair of structures $S_1, S_2 \in \text{STRUCT}(\tau)$, if:

- $|S_1| = |S_2|$;
- $R^{S_1} \subseteq R^{S_2}$, for all relation symbols R in τ;
- $C^{S_1} = C^{S_2}$, for all constant symbols C in τ;
- $S_1 \in \Omega$,

then $S_2 \in \Omega$. The problem Ω over the signature τ is *non-trivial* if

- whenever $S \in \Omega$ we have that the relation R^S is non-empty, for every relation symbol R of τ;
- whenever $T \in \text{STRUCT}(\tau) \setminus \Omega$ and S is a sub-structure of T such that for any relation symbol R of τ, or arity a, say, and for any $\mathbf{u} \in |T|^a \setminus |S|^a$, $R^T(\mathbf{u})$ does not hold, we have $S \notin \Omega$;
- Ω is non-empty.

Intuitively, a problem involving digraphs, for example, is monotone if whenever a digraph is a yes-instance of the problem then adding more edges does not change this fact.

The following proof is a generalization of that in [9].

Theorem 2. *Let Ω be some non-trivial monotone problem, let $\omega = \omega_1\omega_2 \ldots \omega_t$ be some string over the alphabet $\{\exists, \forall\} \cup \{\Omega_k, \neg\Omega_k : k$ a positive integer$\}$ and let $\mathcal{A}$ and $\mathcal{B}$ be two structures over some signature τ. The following are equivalent.*

(*i*) *For any sentence $\psi \in \Omega(\omega)$ over τ, if $\mathcal{A} \models \psi$ then $\mathcal{B} \models \psi$.*
(*ii*) *The Duplicator has a winning strategy in the $\Omega(\omega)$-game on $\mathcal{A}$ and $\mathcal{B}$.*

These games have been used to exhibit a proper hierarchy within the logic TC*[FO] [9]; to examine extensions of $\mathcal{L}^{\omega}_{\infty\omega}$ by families of generalized quantifiers [12]; and to compare problems with respect to logical translations [6].

Yet another role of monotonicity in finite model theory arises in the study of databases, where a database is regarded as being a finite relational structure. Datalog is a database query language [23], with a Datalog program being a collection of rules of the form

$$\alpha_0 \text{ :- } \alpha_1, \ldots, \alpha_k,$$

where α_0 is the *head* and $\alpha_1, \ldots, \alpha_k$ is the *body* of the rule. Each α_i is an atomic formula involving some relation symbol which may be one of the database relations, an *extensional* relation symbol, or alternatively an *intensional* relation symbol: all heads of rules involve intensional relation symbols. Initially, all intensional relations are empty and rules are applied, if allowed according to the input structure (which defines the extensional relations), so as to add tuples to the intensional relations. When no new tuples can be added then the resulting intensional relation is the *query* computed by the program.

Note that if $S_1 \subseteq S_2$, for some structures S_1 and S_2 over the same relational signature, then the query computed by some Datalog program on input S_1 is contained in the query computed by some Datalog program on input S_2; that is, the query is *preserved under extensions*. In database theory, a query is said to be *monotone* if it is preserved under extensions and one-to-one homomorphisms: equivalently, the query computed does not decrease by adding elements to the domain of the input structure or tuples to the database relations. Polynomial-time monotone queries have been shown not to be expressible in Datalog and its variants [1].

4 More Roles for Monotonicity

Our first example of a less well-known role for monotonicity in finite model theory involves an examination of the anomaly arising in Theorem 1. Whereas the logic HP*[FO$_s$] captures NP, the logic 3COL*[FO$_s$] captures L^{NP}, and it is widely believed that NP $\neq$ co-NP, and so NP $\neq$ L^{NP}. Might there be some structural reason for this apparent divergence as, after all, HP and 3COL are merely "typical" NP-complete problems (in the traditional sense)? It had been suggested that the fact that HP is monotone (in the sense defined immediately prior to Theorem 2) whereas 3COL isn't might play some sort of role. (The following results are given in more detail in [21].)

Let us make a very preliminary, working conjecture.

Working Conjecture. Let Ω be some problem that is complete for NP via first-order translations. Then (assuming NP $\neq$ co-NP) $\Omega^*[\mathrm{FO}_s] = \mathrm{NP}$ if and only if Ω is monotone.

Proposition 3. *Let Ω be any monotone problem in NP. Then $\Omega^*[FO_s] \subseteq NP$.*

Proof. For simplicity, suppose that $\sigma(\Omega)$ consists of a unary relation symbol U (the general case proceeds similarly), and let Ω' be some problem described by a sentence of $\Omega^*[\mathrm{FO}_s]$ of the form $\Omega[\lambda\mathbf{x}\psi(\mathbf{x})]$, where the problem described by $\psi(\mathbf{x})$ can be solved in NP and $|\mathbf{x}| = k$. Consider the following NP algorithm on input $S \in \mathrm{STRUCT}(\sigma(\Omega'))$:

given $S \in \mathrm{STRUCT}(\sigma(\Omega'))$, guess a set of tuples $T \subseteq |S|^k$;
verify that $\psi^S(\mathbf{u})$ holds, for each $\mathbf{u} \in T$;
if the structure $T \in \mathrm{STRUCT}(\langle U \rangle)$ of size $|S|^k$ is in Ω then accept, otherwise reject.

The above algorithm accepts Ω' and the result follows by induction.

The above proposition proves "one half" of our working conjecture. Let Ω_1 and Ω_2 be two problems over the same signature. Then Ω_1 is a *finite variation* of Ω_2 if $\Omega_1 \cup \Omega_2 \backslash \Omega_1 \cap \Omega_2$ is finite. We write $\Omega|^{>N}$ to denote those structures of Ω of size greater than N.

Proposition 4. *Let Ω be a finite variation of HP. Then Ω is complete for NP via first-order translations.*

Proof. Let N be such that $\Omega|^{>N} = \mathrm{HP}|^{>N}$ and let m be such that $2^m \geq N$. By Theorem 1, any problem $\Omega' \in \mathrm{NP}$ can be described by a sentence of the form $\mathrm{HP}[\lambda\mathbf{x},\mathbf{y}\psi(\mathbf{x},\mathbf{y})]$, where $\mathbf{x}$ and $\mathbf{y}$ are k-tuples of variables and ψ is a first-order formula. If $k > m$ then Ω' is also described by the sentence $\Omega[\lambda\mathbf{x},\mathbf{y}\psi(\mathbf{x},\mathbf{y})]$. Otherwise, set $j = m - k + 1$ and let $\mathbf{u}$ and $\mathbf{v}$ be j-tuples of new variables (with all variables distinct). Define $\psi'(\mathbf{u},\mathbf{x},\mathbf{v},\mathbf{y})$ as

$$(\mathbf{u} = \mathbf{0} \wedge \mathbf{v} = \mathbf{0} \wedge \psi(\mathbf{x},\mathbf{y})) \vee (\mathbf{u} \neq \mathbf{0} \wedge \mathbf{v} \neq \mathbf{0} \wedge \mathbf{x} \neq \mathbf{y})$$
$$\vee(\mathbf{u} = \mathbf{0} \wedge \mathbf{v} = \mathbf{max} \wedge \mathbf{x} = \mathbf{max} \wedge \mathbf{y} = \mathbf{0})$$

(our shorthand is obvious). The problem Ω' is described by $\mathrm{HP}[\lambda(\mathbf{u},\mathbf{x}),(\mathbf{v},\mathbf{y})\psi']$ and so also by the sentence $\Omega[\lambda(\mathbf{u},\mathbf{x}),(\mathbf{v},\mathbf{y})\psi']$. The result follows as $\Omega \in \mathrm{NP}$.

Proposition 5. *Let Ω be a finite variation of HP. Then $\Omega^*[FO_s] = \Omega^1[FO_s] = NP$.*

Proof. By Proposition 4, $\mathrm{NP} = \Omega^1[\mathrm{FO}_s] \subseteq \Omega^*[\mathrm{FO}_s]$. As our induction hypothesis, suppose that $\Omega^i[\mathrm{FO}_s] = \mathrm{NP}$, for some $i \geq 1$ (the base case holds). Let N be such that $\Omega|^{>N} = \mathrm{HP}|^{>N}$ and suppose that Ω' is some problem describable by a sentence of the form $\Omega[\lambda\mathbf{x},\mathbf{y}\psi(\mathbf{x},\mathbf{y})]$, where $\mathbf{x}$ and $\mathbf{y}$ are k-tuples of variables and ψ is a formula of $\Omega^i[\mathrm{FO}_s]$.

Let $S \in \mathrm{STRUCT}(\tau(\Omega'))$ be of size n. Denote by $\sigma(S)$ the τ_2-translation of S with respect to $\{\psi(\mathbf{x},\mathbf{y})\}$ (and so $\sigma(S)$ is a structure over τ_2 of size n^k). Then $S \models \Omega[\lambda\mathbf{x},\mathbf{y}\psi(\mathbf{x},\mathbf{y})]$ if and only if

$$(n^k > N \text{ and } S \models \mathrm{HP}[\lambda\mathbf{x},\mathbf{y}\psi(\mathbf{x},\mathbf{y})]) \text{ or } (n^k \leq N \text{ and } \sigma(S) \in \Omega).$$

However, by the induction hypothesis and Proposition 3, whether $S \models \mathrm{HP}[\lambda\mathbf{x},\mathbf{y}\psi(\mathbf{x},\mathbf{y})]$ can be verified in NP and, if $n^k \leq N$, whether $\sigma(S) \in \Omega$ can be verified by checking to see if $\sigma(S)$ appears on a fixed list of structures of finite length (independent of n). Consequently, $\Omega^{i+1}[\mathrm{FO}_s] \subseteq \mathrm{NP}$ and the result follows by induction.

Corollary 6. *There exists a problem Ω such that:*

(i) $\Omega^*[FO_s] = \Omega^1[FO_s] = NP$;
(ii) *Ω is complete for NP via first-order translations;*
(iii) *Ω is not monotone.*

Consequently, the Working Conjecture is false.

Proof. Let Ω be any finite variation of HP that is not monotone. The result follows immediately by Propositions 4 and 5.

Note that the counter-examples to the Working Conjecture constructed above are somewhat artificial for, in practice, if a problem is not monotone then there are witnesses as to this non-monotonicity of almost all sizes. For example, consider 3COL. The pair of graphs with no edges and all edges, of any size > 3, are witnesses as to the non-monotonicity of 3COL (for the former can be 3-coloured whilst the latter can't). Also, these witnesses are usually describable in a "uniform" way (as the pairs of witnesses for 3COL above can be).

Whilst our Working Conjecture holds for everyday NP-complete problems (under the assumption that NP $\neq$ co-NP), which tend to be monotone or "uniformly" non-monotone, it might be the case that there exists a non-monotone NP-complete (via first-order translations) problem Ω that is not a finite variation of a monotone NP-complete problem but that $\Omega^*[\mathrm{FO}_s] = \mathrm{NP}$. We refer the reader to [4] for more details regarding non-uniform complexity classes.

Theorem 7. *Let Ω be a non-monotone problem that is non-monotone on structures of size $n \in \{n_0, n_1, \ldots, n_i, \ldots\} \subseteq \mathbf{N}$, where $n_0 < n_1 < \ldots < n_i < \ldots$, and complete for NP via first-order translations. If $\Omega^*[FO_s] = NP$ then there exists a nondeterministic polynomial-time Turing machine M and, for each strictly increasing polynomial p, a polynomial-size advice language A_p such that M with advice A_p accepts the problem X_p defined as*

$$\{S \in STRUCT(\tau_2) : |S| = n, n \in [n_i - p(n), n_i] \text{ for some } i, \text{ and } S \notin HP\}.$$

Proof. Let p be a strictly increasing polynomial and, for simplicity, let us assume that $\tau(\Omega)$ consists of the unary relation symbol U (the general case proceeds similarly). Let Φ be a sentence of $\Omega^1[\mathrm{FO}_s(\tau_2)]$ describing HP (recall, $\tau_2 = \langle E\rangle$). Fix n_i, and let S_{i1} and S_{i2} be structures over $\tau(\Omega)$ of size n_i such that $S_{i1} \subseteq S_{i2}, S_{i1} \in \Omega$, and $S_{i2} \notin \Omega$. Define the sentence Ψ as

$$\Omega[\lambda y(R_1(y) \vee (\Phi \wedge R_2(y))]$$

where R_1 and R_2 are two new unary relation symbols. Then for any $S \in \mathrm{STRUCT}(\tau_2)$ of size n_i, the structure $S' \in \mathrm{STRUCT}(\tau')$ of size n_i, where $\tau' = \langle E, R_1, R_2\rangle$, defined via:

$E^{S'}(x, y)$ holds if and only if $E^S(x, y)$ holds;
$R_1^{S'}(x)$ holds if and only if $U^{S_{i1}}(x)$ holds;
$R_2^{S'}(x)$ holds if and only if $U^{S_{i2}}(x)$ holds,

is such that $S' \models \Psi$ if and only if $S \notin \mathrm{HP}$.

On input to a Turing machine, a structure $S \in \mathrm{STRUCT}(\tau_2)$ of size n is encoded as $e_{\tau_2}(S)$, a string of length n^2; where for any structure S of size n over any signature τ the string $e_\tau(S)$ of 0's and 1's is obtained by concatenating strings of length n^k describing relations of arity k (0 stands for *False* and 1 stands for *True*). Let the string ω_m, where $m = n_i^2$, of 0's and 1's be an encoding of the pair of structures (S_{i1}, S_{i2}), prefixed with a 0: note that there exists a polynomial q such that $|\omega_m| \leq q(n_i)$ for all $i \in \mathbf{N}$.

Consider some $n \in [n_i - p(n), n_i)$ for some i. Let ω_m, where $m = n^2$, be a string of 0's and 1's encoding the triple (n_i, S_{i1}, S_{i2}), all prefixed with a 1. Finally, if $j \in \mathbf{N}$ is not a perfect square then define ω_j to be the empty string, and set $A_p = \{\omega_k : k \in \mathbf{N}\}$.

Let M be a nondeterministic polynomial-time Turing machine, using a polynomial-size advice language A_p, which works as follows:

on input (ω, ω_n), where $|\omega| = n$, M rejects the input if ω_n is the empty string;
if the first digit of ω_n is a 0 then $n = n_i^2$, for some i, and M evaluates the sentence Ψ for the structure $S' \in \mathrm{STRUCT}(\tau')$ of size n_i obtained, as above, from the structure S, where $e_{\tau_2}(S) = \omega$, and the structures S_{i1} and S_{i2}, encoded in ω_n: M accepts if $S \models \Psi$;
if the first digit of ω_n is a 1 then $n = m^2$, for some $m \notin \{n_0, n_1, \ldots, n_i, \ldots\}$: M determines, from ω_n, the n_i such that $n > n_i - p(n)$ and then builds a structure $S_0 \in \mathrm{STRUCT}(\tau_2)$ of size n_i such that $S_0 \in \mathrm{HP}$ if and only if $S \in \mathrm{HP}$, where S is such that $e_{\tau_2}(S) = \omega$: M then evaluates the sentence Ψ for the structure $S' \in \mathrm{STRUCT}(\tau')$ of size n_i obtained, as above, from the structure S_0 and the structures S_{i1} and S_{i2}, encoded in ω_n: M accepts if $S' \models \Psi$.

With advice A_p, M runs in nondeterministic polynomial-time and accepts the problem X_p.

Let Ω be a non-monotone problem that is non-monotone on structures of size $n \in \{n_0, n_1, \ldots, n_i, \ldots\} \subseteq \mathbf{N}$, where $n_0 < n_1 < \ldots < n_i < \ldots$ and where $n_i - p(n_i - 1) < n_{i-1}$ for all i and for some fixed polynomial p. Then we say that Ω is *polynomially-often non-monotone*. The following result is an immediate corollary of Theorem 7. (Here, $\mathcal{NP}$ is non-uniform NP: see [4].)

Corollary 8. *Let Ω be a polynomially-often non-monotone problem that is complete for NP via first-order translations. If $\Omega^*[FO_s] = NP$ then $NP \subseteq co\text{-}\mathcal{NP}$.*

Corollary 8 tells us that if an NP-complete problem (via first-order translations) Ω is non-monotone, but the structure sizes where Ω is non-monotone are "not too far apart", then it is unlikely that $\Omega^*[\mathrm{FO}_s] = \mathrm{NP}$. Note that there exist NP-complete problems (via first-order translations) Ω which are non-monotone but not polynomially-often non-monotone: one such is a problem over τ_2 that is non-monotone on structures whose size is an exponential tower of 3's (that is $3, 3^3, 3^{3^3}, \ldots$) but is identical with HP on structures of other sizes (to see this, note that given some problem $\Omega \in$ NP there is a first-order translation of even arity from Ω to HP).

Our second less well-known role for monotonicity in finite model theory involves the logical capture of complexity classes of monotone problems, defined with respect to some appropriate model of computation. In particular, we focus on the class of monotone NP problems. (The following results are given in more detail in [20].)

We now briefly explain how the usual nondeterministic Turing machine model of computation can be adapted so that the input can be randomly accessed. Our adapted model is similar to those in [5, 14] although it is designed to work on inputs in the form of finite structures as opposed to strings of symbols.

A *Random-Access Turing machine* (*RAT*) M over a signature τ consisting entirely of relation symbols is essentially a (one work tape) nondeterministic Turing machine with its input tape and head replaced by oracle tapes and heads, with an oracle tape and head for each relation symbol of τ. The oracle tapes provide M with a means for accessing the input structure which is always the oracle. Let $S \in$ STRUCT(τ) be of size n. We assume that when S is input to M, the binary representation of n is initially written in the first $\lfloor \log n \rfloor + 1$ cells of the work tape (and so M knows the size of its input structure). Let R be some relation symbol of τ of arity a. On input S, if M needs to know whether $R^S(x_1, x_2, \ldots, x_a)$ holds, where $x_1, x_2, \ldots, x_a \in |S|$, then M writes the binary representation of $(x_1, x_2, \ldots, x_a)$ on the oracle tape corresponding to R. The oracle answers "yes" if and only if $R^S(x_1, x_2, \ldots, x_a)$ holds. If ever the queried string on the oracle tape corresponding to R does not correspond to a tuple $(x_1, x_2, \ldots, x_a)$ then the computation crashes (that is, halts without accepting). The input S is accepted if and only if there is some computation of M on S leading to the accepting state. Consequently, the problem over τ accepted by M consists of all those structures of STRUCT(τ) accepted by M. We define the time taken and space used by a RAT as usual (for space, we count the number of work tape cells used). The complexity class $\mathrm{NP}_{\mathrm{RAT}}(\tau)$ is the class of all problems over τ accepted by a polynomial time RAT over τ: it is easy to see that $NP_{RAT} = NP$, where NP_{RAT} is the class of all problems accepted by a polynomial time RAT (over some signature consisting entirely of relation symbols).

We are more concerned with when a RAT M is restricted to operate *conjunctively*. By this we mean that if ever the answer to some oracle query is "no" then the computation crashes. If M is a polynomial time conjunctive RAT (C-RAT) then there is an equivalent polynomial time C-RAT M' with the property that every computation is split into two parts. The first part consists of M' simulating M except that no oracle queries are made: the queries are actually computed and stored on the work tape, with M' assuming that all the answers to the (unmade) oracle queries are "yes". If the simulation of M halts in the accepting state then in the second part of its computation, M' queries the oracle for each of the remembered oracle queries and accepts if and only if all the oracle answers are "yes" (we may assume that the simulation does indeed halt as M is a polynomial time C-RAT and can be augmented with a clock). Consequently, throughout the paper we assume that any polynomial time C-RAT computes all of its oracle queries before actually making any. The complexity class $NP_{C\text{-}RAT}(\tau)$ is the class of all problems over τ accepted by a polynomial time C-RAT over τ, with the complexity class $NP_{C\text{-}RAT}$ as expected.

It is easy to see that $NP_{C\text{-}RAT}$ is a complexity class similar to the class of problems defined by the sentences of existential monadic second-order logic (see [2]), for $NP_{C\text{-}RAT}$ contains many ("hard") NP-complete (via logspace reductions) problems but does not contain some ("easy") problems solvable in P (although existential monadic second-order logic and $NP_{C\text{-}RAT}$ are incomparable). Furthermore, it is easy to prove the following.

Theorem 9. *The complexity class $NP_{C\text{-}RAT}$ consists of all those monotone problems in NP.*

We begin by considering a logical encoding of the path systems accessibility problem ([8]), the well-known decision problem that is defined as follows: an instance (that is, a *path system*) of size n is a set V of n vertices, a relation $R \subseteq V \times V \times V$, a *source* $u \in V$, and a *sink* $v \in V$ (different from u), and a yes-instance is an instance where the sink is accessible, with a vertex w being *accessible* if w is the source or if $R(x, y, w)$ holds for some accessible vertices x and y. In deducing that a vertex w is accessible by exhibiting accessible vertices x and y such that $R(x, y, w)$ holds, we sometimes say that we have *applied the rule* (x, y, w): we remark that x and y need not be distinct. We encode this decision problem as the problem PS over the vocabulary $\tau_3 = \langle R \rangle$, where R is a relation symbol of arity 3, in the obvious manner with 0 (resp. max) representing the source (resp. sink).

Remark. Until further notice, constant symbols are allowed in our signatures except where otherwise stated.

Let $\phi \in \mathrm{FO}_s(\tau)$, for some signature τ, be of the form

$$\bigvee\{\alpha_i \wedge \beta_i : i \in I\}$$

for some finite index set I, where:

- each α_i is a conjunction of the logical atomic relations s, $=$, and their negations, and no symbol of τ appears in any α_i;
- β_i is atomic or negated atomic;
- if $i \neq j$ then α_i and α_j are mutually exclusive.

Then ϕ is a *projective formula*.

Theorem 10. *Every formula Φ of the logic $PS^*[FO_s]$ is equivalent to one of the form*

$$PS[\lambda \mathbf{xyz}\psi]$$

where $\mathbf{x}$, $\mathbf{y}$, *and* $\mathbf{z}$ *are k-tuples of variables, for some k, and where ψ is a projective formula.*

The proof of the above theorem is omitted and proceeds by quantifier elimination (as in, for example, [11, 17]). The following corollary can then be easily proven.

Corollary 11. $(\pm PS)^*[FO_s] = PS^*[FO_s] = PS^1[FO_s] = P$ *and PS is complete for P via projection translations.*

In fact, we can say even more. Let a projective formula be *monotone* if all occurrences of relation symbols of the underlying signature are atomic (and *not* negated atomic). Also, let the logic $\mathrm{PS}^*[\mathrm{FO}_s^+]$ be the fragment of $\mathrm{PS}^*[\mathrm{FO}_s]$ where all occurrences of any symbol of the underlying signature appear positively, i.e., not within the scope of any negation sign.

Corollary 12. *The problem PS is complete for $PS^*[FO_s^+]$ via monotone projection translations.*

Problems describable by sentences of the logic $\mathrm{PS}^*[\mathrm{FO}_s^+]$ can be recognized by sequences of monotone circuits of polynomial size: in fact, they can be recognized by uniform sequences of such circuits (under a suitable uniformity constraint such as logspace computability). Consequently, by Razborov's result that there are monotone problems in P that cannot be recognized by sequences of monotone circuits of polynomial size [13], we can say that there are monotone problems in P that are not expressible by sentences of $\mathrm{PS}^*[\mathrm{FO}_s^+]$. It would be interesting to investigate the complexity class captured by the logic $\mathrm{PS}^*[\mathrm{FO}_s^+]$: for example, does it coincide with the class of problems recognized by uniform sequences of monotone circuits of polynomial size?

We now use the above results to logically characterize the complexity class $\mathrm{NP}_{\text{C-RAT}}$.

Remark. Until further notice, signatures consist entirely of relation symbols except where otherwise stated.

Theorem 13. *Let Ω be some problem over the signature τ which can be accepted by some polynomial time C-RAT (over τ). Then Ω can be described by some sentence of the form*

$$\exists G\Phi$$

where G is a relation symbol not in τ and where $\Phi \in PS^*[FO_s(\tau \cup \langle G\rangle)]$ with all occurrences of any symbol of τ positive.

Proof. Let Ω be some problem over the signature $\tau = \langle R_1, R_2, \ldots, R_r\rangle$, where each R_i is a relation symbol of arity a_i, which can be accepted by some C-RAT M over τ of time complexity n^k. Let G be a new relation symbol of arity k and let $D_1, D_2, \ldots, D_{a_i}$ be constant symbols, for some $i \in \{1, 2, \ldots, r\}$. Define the signature $\tau_i' = \langle G, D_1, D_2, \ldots, D_{a_i}\rangle$ and the problem Ω_i' over τ_i' as

> $\{S' \in \mathrm{STRUCT}(\tau_i')$: on input $S \in \mathrm{STRUCT}(\tau)$, where S is any structure of size $|S'|$, the C-RAT M with nondeterministic guesses given by $G^{S'}$ queries whether $R_i^S(D_1^{S'}, D_2^{S'}, \ldots, D_{a_i}^{S'})$ holds$\}$

(recall that a C-RAT defers all oracle queries to the end of the computation). Clearly, Ω_i' can be solved in deterministic polynomial time and so by Corollary 11 there is a sentence of $\mathrm{PS}^*[\mathrm{FO}_s(\tau_i')]$ of the form

$$\mathrm{PS}[\lambda \mathbf{xyz}\psi_i'(\mathbf{x}, \mathbf{y}, \mathbf{z})]$$

describing Ω_i', where $\mathbf{x}$, $\mathbf{y}$, and $\mathbf{z}$ are tuples of variables of the same length and $\psi_i'(\mathbf{x}, \mathbf{y}, \mathbf{z})$ is a projective formula (remember, we can have constant symbols in our vocabularies for Corollary 11). Replace any occurrence of the constant symbol D_j in ψ_i' by the (new) variable u_j, for each j, to get the formula ψ_i (over $\langle G\rangle$).

Finally, let Ω' be the problem over $\tau' = \langle G\rangle$ defined as

> $\{S' \in \mathrm{STRUCT}(\tau')$: on input $S \in \mathrm{STRUCT}(\tau)$, where S is any structure of size $|S'|$, the C-RAT M with nondeterministic guesses given by $G^{S'}$ accepts S so long as all the answers to the oracle queries are "yes"$\}$.

Then Ω' can be solved in deterministic polynomial time and so by Corollary 11 there is a sentence of $\mathrm{PS}^*[\mathrm{FO}_s(\tau')]$ of the form

$$\mathrm{PS}[\lambda \mathbf{xyz}\phi(\mathbf{x}, \mathbf{y}, \mathbf{z})]$$

describing Ω', where $\mathbf{x}$, $\mathbf{y}$, and $\mathbf{z}$ are tuples of variables of the same length and $\phi(\mathbf{x}, \mathbf{y}, \mathbf{z})$ is a projective formula.

Consequently, for any $S \in \mathrm{STRUCT}(\tau)$, $S \in \Omega$ if and only if

$$S \models \exists G(\mathrm{PS}[\lambda \mathbf{xyz}\phi(\mathbf{x}, \mathbf{y}, \mathbf{z})] \wedge \forall \mathbf{u}(\mathrm{PS}[\lambda \mathbf{xyz}\psi_1(\mathbf{x}, \mathbf{y}, \mathbf{z}; \mathbf{u}_1)] \Rightarrow R_1(\mathbf{u}_1) \\ \wedge \ldots \wedge \mathrm{PS}[\lambda \mathbf{xyz}\psi_r(\mathbf{x}, \mathbf{y}, \mathbf{z}; \mathbf{u}_r)] \Rightarrow R_r(\mathbf{u}_r)))$$

where $\mathbf{u} = (u_1, u_2, \ldots, u_t)$, with $t = max\{a_i : i = 1, 2, \ldots, r\}$, and $\mathbf{u}_i = (u_1, u_2, \ldots, u_{a_i})$, for $i = 1, 2, \ldots, r$ (we may clearly assume that the lengths of the tuples $\mathbf{x}$, $\mathbf{y}$, and $\mathbf{z}$ in each formula ψ_i or ϕ are the same).

It should be clear from the proof of Theorem 13 why we sometimes need constant symbols in our vocabularies.

Definition 14. The problem NES (standing for Non-Empty Satisfiability) consists of all those structures over the signature $\tau_{2,2} = \langle P, N\rangle$, where P and N are binary relation symbols, encoding satisfiable boolean formulae in conjunctive normal form where every clause is non-empty, via: given $S \in \text{STRUCT}(\tau_{2,2})$, for every $i, j \in |S|$, clause j contains the literal X_i (resp. $\neg X_i$) if and only if $P^S(i,j)$ (resp. $N^S(i,j)$) holds.

Definition 15. A *labelled path system* P of size n is a path system $(V, R, \text{source}, \text{sink})$ of size n where if $R(x,y,z)$ holds then the rule (x,y,z) has an associated boolean literal from the set $\{X_0, \neg X_0, X_1, \neg X_1, \ldots, X_{n-1}, \neg X_{n-1}, \mathit{True}\}$. The sink is accessible in this labelled path system if there is a truth assignment t on the boolean variables $\{X_0, X_1, \ldots, X_{n-1}\}$ such that the sink is accessible in the path system obtained from P by retaining only those rules whose associated literal is set at *True* under t. Labelled path systems accessibility is the decision problem of deciding whether the sink is accessible in some labelled path system. We encode this decision problem as the problem LPS over the signature $\tau = \langle L_p, L_n, T\rangle$, where L_p and L_n are relation symbols of arity 4 and T is a relation symbol of arity 3, by stipulating that for any labelled path system:

$L_p(x,y,z,u)$ holds if and only if the rule (x,y,z) is labelled with X_u;
$L_n(x,y,z,u)$ holds if and only if the rule (x,y,z) is labelled with $\neg X_u$;
$T(x,y,z)$ holds if and only if the rule (x,y,z) is labelled with *True*

(notice that not every structure over τ corresponds to a labelled path system but we can describe exactly those structures that do using a first-order formula: this is analogous to the situation in traditional complexity theory where not every string is the encoding of some instance of a decision problem). Of course, if $S \in \text{STRUCT}(\tau)$ does not encode a labelled path system then $S \notin \text{LPS}$.

Theorem 16. *Let Ω be some problem over the signature τ which can be described by some sentence of the form*

$$\exists G \Phi$$

where G is a relation symbol not in τ and where $\Phi \in PS^*[FO_s(\tau \cup \langle G\rangle)]$ with all occurrences of any symbol of τ positive. Then Ω can be described by a sentence of $NES^*(FO_s^+(\tau))$ of the form

$$NES[\lambda \mathbf{xy}\psi_p, \mathbf{xy}\psi_n]$$

where $\mathbf{x}$ and $\mathbf{y}$ are k-tuples of variables, for some k, and ψ_p and ψ_n are monotone projective formulae (over τ).

Proof. (Sketch) We use Corollary 12 to obtain a normal form for the formula Φ in the statement of the theorem, and then use the technique first established in [16] to move from a Fagin-style description to an Immerman-style description. It turns out that the problem Ω of the theorem can then be described by a sentence Φ' of the form:

$$\mathrm{LPS}[\lambda \mathbf{xyzu}\phi_1(\mathbf{x},\mathbf{y},\mathbf{z},\mathbf{u}), \mathbf{xyzu}\phi_2(\mathbf{x},\mathbf{y},\mathbf{z},\mathbf{u}), \mathbf{xyz}\phi_3(\mathbf{x},\mathbf{y},\mathbf{z})],$$

where ϕ_1, ϕ_2 and ϕ_3 are quantifier-free. The result follows by showing that the sentence Φ' is equivalent to one of the required form.

Corollary 17. *Let Ω be some problem over τ. Then $\Omega \in NP_{C\text{-}RAT}$ if and only if Ω can be described by a sentence of $NES^*[FO_s^+(\tau)]$ of the form*

$$NES[\lambda \mathbf{xy}\psi_p, \mathbf{xy}\psi_n]$$

where $\mathbf{x}$ and $\mathbf{y}$ are k-tuples of variables, for some k, and ψ_p and ψ_n are monotone projective formulae.

Proof. Immediate from Theorems 13 and 16.

Corollary 18. $NP_{C\text{-}RAT} = NES^*[FO_s^+]$.

Proof. Immediate from Corollary 17.

We need the version (NES) of the satisfiability problem where every clause is known to contain at least one literal so that the problem can be solved by a polynomial time C-RAT.

Corollary 19. *For any signature τ, $NP_{C\text{-}RAT}(\tau)$ is captured by the sub-logic of existential second-order logic with sentences of the form*

$$\exists T\phi$$

where T is some relation symbol not in τ and ϕ is a first-order sentence over $\tau \cup \langle T\rangle$ with all occurrences of any relation symbol of τ positive.

Proof. By Corollary 17, any problem of $\mathrm{NP}_{\mathrm{C\text{-}RAT}}(\tau)$ can be described by a sentence of the form

$$\exists T\forall\mathbf{y}\exists\mathbf{x}((\psi_p(\mathbf{x},\mathbf{y}) \wedge T(\mathbf{x})) \vee (\psi_n(\mathbf{x},\mathbf{y}) \wedge \neg T(\mathbf{x})))$$

where $\mathbf{x}$ and $\mathbf{y}$ are k-tuples of variables, for some k, ψ_p and ψ_n are monotone projective formulae, and T is a relation symbol of arity k ($\forall\mathbf{y}$, for example, is shorthand for $\forall y_1 \forall y_2 \ldots \forall y_k$ if $\mathbf{y} = (y_1, y_2, \ldots, y_k)$). The converse containment is obvious.

Corollary 19 should be compared with Fagin's result that NP is captured by existential second-order logic [7]. We should add that Corollary 19 could probably have been proven by resorting to Fagin's methods [7]. However, we prefer our proof as it shows how we can move from Immerman-type descriptions to Fagin-type ones, we do not need to drop down to the level of Turing machines, and we can use existing results. Also, we obtain a particularly nice normal form (see the proof of Corollary 19) for formulae of the sub-logic L involved in the statement of Corollary 19. Bearing this normal form in mind, it might be worthwhile looking at the complexity classes captured when we restrict the first-order quantifier prefix in the formulae of the logic L: for example, we might insist that the first-order quantifier prefix consists entirely of universal quantifiers.

5 Conclusion

We have attempted here to state a case for a further, systematic study of monotonicity in finite model theory, particularly with regard to the capture of monotone complexity classes. For example, it would be interesting to capture the class of polynomial-time or logspace solvable monotone problems using some logic: these classes have so far resisted capture by any reasonable model of computation. Any study of monotonicity might go hand in hand with a study of the complexity-theoretic concept of non-uniformity within finite model theory, as hinted at earlier. This is also an area where not much work has been done.

References

1. F. Afrati and S.S. Cosmadakis and M. Yannakakis, On Datalog vs. polynomial time, *Proc. 10th ACM Ann. Symp. on Principles of Database Systems* (1991) 13–23
2. M. Ajtai and R. Fagin, Reachability is harder for directed than for undirected finite graphs, *J. Symbolic Logic* **55** (1990) 113–150
3. M. Ajtai and Y. Gurevich, Monotone versus positive, *J. Assoc. Comput. Mach.* **34** (1987) 1004–1015
4. J.L. Balcázar and J. Díaz and J. Gabarró, *Structural Complexity Theory I*, Springer-Verlag, Berlin (1988)
5. A.K. Chandra and D.C. Kozen and L.J. Stockmeyer, Alternation, *J. Assoc. Comput. Mach.* **28** (1981) 14–133
6. A. Dawar and I.A. Stewart, manuscript
7. R. Fagin, Generalized first-order spectra and polynomial-time recognizable sets, *Complexity of Computation*, SIAM-AMS Proceedings, Vol. 7 (ed. R.M. Karp) (1974) 43–73
8. M.R. Garey and D.S. Johnson, *Computers and Intractability: a Guide to the Theory of NP-Completeness*, W.H. Freeman and Co., San Francisco (1979)
9. E. Grädel, On transitive closure logic, *Lecture Notes in Computer Science Vol.* 626 (1992) 149–163
10. N. Immerman, Relational queries computable in polynomial time, *Inform. Control* **68** (1986) 86–104
11. N. Immerman, Languages that capture complexity classes, *SIAM J. Comput.* **16** (1987) 760–778
12. P.G. Kolaitis and J.A. Väänänen, Generalised quantifiers and pebble games on finite structures, *J. Pure App. Logic*, to appear
13. A.A. Razborov, A lower bound on the monotone network complexity of the logical permanent, *Mat. Zametki* **41** (1987) 598–607 (in Russian; English translation in: *Math. Notes*, **41** (1987), 333–338)
14. W.L. Ruzzo, On uniform circuit complexity, *J. Comput. System Sci.* **22** (1981) 365–383
15. I.A. Stewart, Comparing the expressibility of languages formed using NP-complete operators, *J. Logic Computat.* **1** (1991) 305–330
16. I.A. Stewart, Complete problems involving boolean labelled structures and projection translations, *J. Logic Computat.* **1** (1991) 861–882

17. I.A. Stewart, Using the Hamiltonian path operator to capture NP, *J. Comput. System Sci.* **45** (1992) 127-151
18. I.A. Stewart, Logical characterizations of bounded query classes I: logspace oracle machines, *Fund. Informat.* **18** (1993) 65–92
19. I.A. Stewart, Logical characterizations of bounded query classes II: polynomial-time oracle machines, *Fund. Informat.* **18** (1993) 93–105
20. I.A. Stewart, Logical descriptions of monotone NP problems, *J. Logic Computat.* **4** (1994) 337–357
21. I.A. Stewart, Monotonicity and the expressibility of NP operators, *Math. Logic Quart.* **40** (1994) 132–140
22. I.A. Stewart, On completeness for NP via projection translations, *Math. Systems Theory* **27** (1994) 125–157
23. J.D. Ullman, *Database and Knowledge-Base Systems*, Computer Science Press (1989)
24. M. Vardi, Complexity of relational query languages, *Proc. 14th ACM Ann. Symp. on Theory of Computing* (1982) 137–146

Numbers Defined by Turing Machines

Rudolf Freund[1]

Technische Universität
Wien
Institut für Computersprachen
Resselgasse 3
A-1040 Wien
Austria

Ludwig Staiger[2]

Martin-Luther-Universität
Halle-Wittenberg
Institut für Informatik
Weinbergweg 17
D-06120 Halle (Saale)
Germany

Abstract

We consider three types of Turing machines defining functions on infinite words and investigate some characteristic properties of these types of Turing machine mappings. Using the interpretation of infinite words as the expansions of numbers we obtain three classes of real respectively complex numbers. We prove that the three classes of complex numbers form algebraically closed subfields of the field of complex numbers.

1 Introduction

Turing machines with infinite input and output are a well-known model from recursion and formal language theory for realizing infinite computations. Infinite words over a finite alphabet X of cardinality $r \geq 2$ in a natural way may be interpreted as r-adic expansions of real numbers (see [3], [2], [10]).

In the following section we introduce three different types of Turing machines. In Section 3 the functions realized by these types of machines are investigated; we characterize their input-output-relations using recursive predicates. In Section 4 we show how these types of Turing machines define real functions taking into account the ambiguities resulting from the interpretation of infinite words as expansions of real numbers.

In the fifth section we derive three different classes of real numbers from our types of Turing machines. The classes of numbers are characterized within the arithmetical hierarchy. Finally we show that the three corresponding classes of complex numbers form algebraically closed subfields of the field of complex numbers.

[1]email: rudi@logic.tuwien.ac.at

[2]email: staiger@informatik.uni-halle.de

2 Definitions

We specify only a few notions and notations here; the reader is referred to [5] for other elements of formal language theory, to [4] for the elements of recursion theory, and to [1] for basic notions in the area of interpreting strings as (real) numbers.

For an alphabet X, by X^* we denote the free monoid generated by X under the operation of concatenation, i.e. the set of all finite words over X including the empty word λ. For a word $w \in X^*$ its *length* is denoted by $|w|$. For words $w, v \in X^*$ we define the prefix order $w \sqsubseteq v$ provided w is a prefix of v. A function $\varphi : X^* \to Y^*$ will be referred to as a *sequential function* if φ is monotone with respect to $\sqsubseteq$ (i.e. $w \sqsubseteq v$ implies $\varphi(w) \sqsubseteq \varphi(v)$), and we call φ *nondecreasing* if $w \sqsubseteq v$ implies $|\varphi(w)| \leq |\varphi(v)|$.

By X^ω we denote the set of all infinite words over X; for an infinite word $\xi \in X^\omega$ the prefix of length m, $m \geq 0$, is denoted by ξ/m; moreover, we define

$$\mathbf{pref}(\xi) = \{\xi/m : m \geq 0\} .$$

We consider the following model of Turing machines: The Turing machine M has a semi-infinite read-only input tape, a semi-infinite output tape and a finite number of working tapes. Since we are interested in infinite strings (and infinite computations), we say that M has a valid output if and only if it writes on every tape cell of the output tape and the contents of every tape cell of the output tape are changed only finitely many times (an equivalent formulation is that every tape cell of the output tape is visited only finitely many times).

The Turing machines introduced above allow us to define Turing machine mappings in the following way:

For a non-deterministic Turing machine M with input and output alphabets X and Y, respectively, the mapping $\Phi_M : X^\omega \to Y^\omega$ is defined as follows: We consider the set of all valid outputs for the input ξ. If this set contains exactly one element η, we set $\Phi_M(\xi) = \eta$. Otherwise $\Phi_M(\xi)$ is undefined.

For a deterministic Turing machine M the n-th letter of $\Phi_M(\xi)$, $\Phi_M(\xi) \in Y^\omega$, is the ultimate contents of the n-th cell of the output tape, provided a valid output appears. Otherwise $\Phi_M(\xi)$ is undefined.

We call a deterministic Turing machine *strict* if it never changes a symbol written on the output tape. Hence if $\Phi_M(\xi)$, $\Phi_M(\xi) \in Y^\omega$, exists, whenever M writes a letter $a \in Y$ on the i-th cell of its output tape we know that the i-th letter of $\Phi_M(\xi)$ is a. In the following we therefore can, without loss of generality, assume that M on its output tape only goes from the left to the right. This also shows that $\Phi_M(\xi)$ can only be undefined if M writes on only finitely many cells of its output tape.

We will use the abbreviations TM, DTM and SDTM for (non-deterministic) Turing machines, deterministic Turing machines, and strict deterministic Turing machines, respectively.

For strings of the forms $+w \cdot p$ and $-w \cdot p$ where $w \in (B_r - \{0\}) B_r^* \cup \{0\}$ and $p \in B_r^\omega$, $B_r = \{0, 1, ..., r-1\}$, $r > 1$, by $\mu_r(+w \cdot p)$ (respectively $(-w \cdot p)$) we denote the real number having $+w \cdot p$ (respectively $(-w \cdot p)$) as *r-adic expansion*. For $p = p(1)p(2)...$, $p(i) \in B_r$ for $i \geq 1$, and $p_0 = +w$ or $p_0 = -w$ for some $w \in (B_r - \{0\}) B_r^* \cup \{0\}$ we define $p_0 \cdot p(1)p(2)... \mid 0 := p_0 \cdot$ and $p_0 \cdot p(1)...p(n)p(n+1)... \mid n := p_0 \cdot p(1)...p(n)$ for $n \geq 1$. The expansion of a real number is unique except for rational numbers of the form j/r^i; e. g. for $r^i > j > 0$, j/r^i has a *terminating expansion* of the form $+0 \cdot u(r-1)0^\omega$ and a *non-terminating expansion* of the form $+0 \cdot u0(r-1)^\omega$, where $u \in B_r^*$ and $\mu_r(+0 \cdot u(r-1)0^\omega) = \mu_r(+0 \cdot u0(r-1)^\omega) = j/r^i$. The set of rational numbers is denoted by Q, the set of all rational numbers of the form j/r^i is denoted by $Q_{r,2}$; we should also mention the special case of the rational number $0 \in Q_{r,2}$, which has the two expansions $+0 \cdot 0^\omega$ and $-0 \cdot 0^\omega$. The set of all strings of the form as above of expansions of real numbers at base r is denoted by R_r. The set of natural numbers $\{0, 1, 2, ...\}$ is denoted by N, the set of real numbers by R, and the set of complex numbers by C.

We may regard Turing machines defining mappings $\Phi_M : R_r \rightarrow R_s$ as devices for computing real functions. Due to the ambiguities of the expansions of the rational numbers in $Q_{r,2}$ respectively $Q_{s,2}$, following [1] we call a mapping Φ_M *consistent* provided that for all $x, y \in dom(\Phi_M)$ with $\mu_r(x) = \mu_r(y)$ also $\mu_s(\Phi_M(x)) = \mu_s(\Phi_M(y))$. The real function induced by such a consistent Turing machine mapping Φ_M is denoted by Φ_M^R. Observe that when being interested in Turing machines as representations of real functions, we do not take care of the actions of these Turing machines on inputs from $(B_r \cup \{+, -, \cdot\})^\omega - R_r$ (we can assume that they halt on such inputs, because the feature that an input from $(B_r \cup \{+, -, \cdot\})^\omega$ is not in R_r can easily be detected by any type of Turing machines considered in this paper).

3 Functions Realized by Turing Machines

Mappings defined by DTMs respectively SDTMs allow for the following characterizations using recursive functions:

Theorem 1 *Let* $\Phi : X^\omega \rightarrow Y^\omega$.

1. Φ *is an SDTM-mapping if and only if there is a recursive sequential function* $\varphi : X^* \rightarrow Y^*$ *such that*

$$\Phi(\xi) = \eta \Longleftrightarrow \forall i \; \exists t \; (\eta/i \sqsubseteq \varphi(\xi/t)) .$$

2. Φ *is a DTM-mapping if and only if there is a recursive nondecreasing function* $\varphi : X^* \rightarrow Y^*$ *such that*

$$\Phi(\xi) = \eta \Longleftrightarrow \forall i \; \exists t \; \forall n \; (n \geq t \rightarrow \eta/i \sqsubseteq \varphi(\xi/n)) .$$

Proof.

1. Let Φ be an SDTM-mapping generated by the Turing machine M. Without loss of generality we may assume M to read the whole input tape. Define $\varphi(w)$ as the contents of the output tape which appears up to that moment after which M tries to read on the input tape beyond the right end of the prefix w of the input $\xi \in X^\omega$. Clearly, φ is a recursive function monotone with respect to $\sqsubseteq$ which satisfies the condition of the theorem.

   Conversely, let $\varphi : X^* \to Y^*$ be a recursive sequential function. From a Turing machine M_φ realizing this function we define the Turing machine M_Φ generating Φ as follows: M_Φ successively reads the prefices w_i, $|w_i| = i$, of the input sequence $\xi \in X^\omega$, computes $\varphi(w_i)$ by simulating M_φ on w_i and adds the (possibly empty) suffix which completes the previously computed output $\varphi(w_{i-1})$ to $\varphi(w_i)$ on the contents of the output tape.

2. In case of a DTM-mapping the construction of the function φ from the machine M is just the same (we assume M never to change an already visited cell on the output tape to blank again).

   In order to prove the converse direction we can also refer to the proof given above with the only exception that M_Φ does not only add new letters to the contents of the output tape but also makes the according changes to previously written tape cells.

q. e. d.

Observe that in case of a recursive sequential function φ the corresponding SDTM does not compute a valid output for an input ξ if and only if the length of the $\varphi(w_i)$ becomes constant after a finite number of steps. In the case of a non-decreasing function φ the corresponding value of Φ may also be undefined if the unboundedly increasing values of $\varphi(w_i)$ do not converge to an infinite word.

In [9] it was shown that in case of fully defined (S)DTM-mappings $\Phi : X^\omega \to Y^\omega$ the prefix of quantifiers in the characterizations of Theorem 1 can be expressed with shorter prefices of quantifiers:

Proposition 1 *Let $\Phi : X^\omega \to Y^\omega$ with dom$(\Phi) = X^\omega$.*

1. *Φ is an SDTM-mapping if and only if there is a recursive relation $R \subseteq X^* \times X^*$ such that*

$$w \in \mathbf{pref}(\Phi(\xi)) \Longleftrightarrow \exists t\, ((w, \xi/t) \in R).$$

2. *Φ is a DTM-mapping if and only if there is a recursive relation $R \subseteq X^* \times N \times X^*$ such that*

$$w \in \mathbf{pref}(\Phi(\xi)) \Longleftrightarrow \exists t \forall n\, ((w, t, \xi/n) \in R).$$

It is interesting to note that there also exist other recursive relations that allow us to characterize (S)DTM mappings by different prefices of quantifiers:

Proposition 2 *Let $\Phi : X^\omega \to Y^\omega$ with $dom(\Phi) = X^\omega$.*

1. *Φ is an SDTM-mapping if and only if there is a recursive relation $R' \subseteq X^* \times X^*$ such that*

$$w \in \mathbf{pref}(\Phi(\xi)) \iff \forall m\,(f(w, \xi/m) \in R').$$

2. *Φ is a DTM-mapping if and only if there is a recursive relation $R' \subseteq X^* \times N \times X^*$ such that*

$$w \in \mathbf{pref}(\Phi(\xi)) \iff \forall t \exists n\,((w, t, \xi/n) \in R').$$

Moreover, it was shown in [9] that DTM-mappings are not closed under composition even if they are fully defined. On the other hand we can easily show that the class of SDTM-mappings is closed under composition and that the composition of a DTM-mapping with SDTM-mappings is again a DTM-mapping:

Lemma 1 *Let $\Phi : X^\omega \to Y^\omega$ and $\Psi : Y^\omega \to Z^\omega$.*

1. *If Φ and Ψ are SDTM-mappings, then $\Psi \circ \Phi$ is an SDTM-mapping, too.*
2. *If Φ is an SDTM-mapping and Ψ is a DTM-mapping, then $\Psi \circ \Phi$ is a DTM-mapping, too.*
3. *If Φ is an DTM-mapping and Ψ is an SDTM-mapping, then $\Psi \circ \Phi$ is a DTM-mapping, too.*

Proof.

1. Let M_Φ be an SDTM realizing Φ and let M_Ψ be an SDTM realizing Ψ. Without loss of generality we may assume M_Ψ to read the whole input tape for every input $\xi \in Y^\omega$. Then an SDTM $M_{\Psi\circ\Phi}$ generating $\Psi \circ \Phi$ can easily be constructed from M_Φ and M_Ψ in the following way: The ouput tape of M_Φ is realized on a working tape of $M_{\Psi\circ\Phi}$, which also serves as the input tape for (the simulation of) M_Ψ. The output tape of $M_{\Psi\circ\Phi}$ represents the output tape of M_Ψ. $M_{\Psi\circ\Phi}$ simulates the actions of M_Ψ until M_Ψ needs a new letter from its input tape. Then $M_{\Psi\circ\Phi}$ simulates the actions of M_Φ until M_Φ writes a new letter on its output tape, which then can be used to continue the simulation of M_Ψ by $M_{\Psi\circ\Phi}$ etc.
2. Let M_Φ be an SDTM realizing Φ and let M_Ψ be a DTM realizing Ψ; then we can use exactly the same construction for $M_{\Psi\circ\Phi}$ as above.

3. Let M_Φ be a DTM realizing Φ and let M_Ψ be an SDTM realizing Ψ. Without loss of generality we may assume M_Ψ to read the whole input tape for every input $\xi \in Y^\omega$. Then a DTM $M_{\Psi\circ\Phi}$ generating $\Psi \circ \Phi$ can be constructed from M_Φ and M_Ψ in the following way: The ouput tape of M_Φ is realized on a working tape T_Φ of $M_{\Psi\circ\Phi}$, which also serves as the input tape for (the simulation of) M_Ψ. The output tape of M_Ψ is realized on another working tape T_Ψ of $M_{\Psi\circ\Phi}$, and we also use another working tape $T_{\Psi\circ\Phi}$ for storing the already computed contents of the ouput tape of $M_{\Psi\circ\Phi}$.

   $M_{\Psi\circ\Phi}$ simulates the actions of M_Φ until M_Φ makes a change on its output tape (i.e. adds a new letter or goes to the left with possibly changing a previously written symbol on T_Φ). In order to ensure that oscillations of M_Φ on its output tape are carried over to the output tape of $M_{\Psi\circ\Phi}$, now $M_{\Psi\circ\Phi}$ moves its head on its ouput tape to the same position as its head on T_Φ and changes the contents of this cell of its output tape.

   Then $M_{\Psi\circ\Phi}$ simulates the actions of M_Ψ on this finite part of the input tape for M_Ψ (the end of this finite part is the current position of the head of $M_{\Psi\circ\Phi}$ on T_Φ) until M_Ψ would need a new letter from its input tape that cannot be provided any more by this finite part already computed on T_Φ. Then $M_{\Psi\circ\Phi}$ checks the contents of its output tape: $M_{\Psi\circ\Phi}$ compares (from the left to the right) the symbols on T_Ψ with the corresponding symbols on $T_{\Psi\circ\Phi}$. When for the first time the corresponding symbols do not match (or when the end of the computed symbols on T_Ψ is reached) $M_{\Psi\circ\Phi}$ goes back to the left to the corresponding position on its output tape and changes its contents as well as the contents of $T_{\Psi\circ\Phi}$ in such a way that they coincide with the contents of T_Ψ.

   After that, $M_{\Psi\circ\Phi}$ continues with the simulation of M_Φ etc.

q. e. d.

Next, we characterize the domains of DTM-mappings. For this aim we introduce some classes of the arithmetical hierarchy:

We say that a set $F \subseteq X^\omega$ is Π_j*-definable* provided there is a recursive relation $R \subseteq (N)^{j-1} \times X^*$ such that

$$\xi \in F \Leftrightarrow \forall i_1 \exists i_2 \ldots \mathbf{Q}_j i_j \, ((i_1, i_2, \ldots, i_{j-1}, \xi/i_j) \in R).$$

In a similar way, a language $W \subseteq X^*$ is called Π_j*-definable* (or a Π_j-set) if and only if

$$w \in W \Leftrightarrow \forall i_1 \exists i_2 \ldots \mathbf{Q} i_j((i_1, i_2, \ldots, i_j, w) \in R)$$

for some recursive relation $R \subseteq N^j \times X^*$. Σ_j-definable sets are the complements of Π_j-definable sets.

The following characterization of Π_1- and Π_2-definable subsets of X^ω can be found in [6].

Proposition 3 *Let $F \subseteq X^\omega$.*

1. *F is Π_1-definable if and only if there is a language W such that $F = \{\xi : \mathbf{pref}(\xi) \subseteq W\}$ and the language $\bigcup_{\xi \in F} \mathbf{pref}(\xi)$ is the complement of a recursively enumerable language.*

2. *F is Π_1-definable if and only if there is a recursive language W such that $F = \{\xi : \mathbf{pref}(\xi) \subseteq W\}$.*

3. *F is Π_2-definable if and only if there is a recursive language V such that $F = \{\xi : \mathbf{pref}(\xi) \cap V$ is infinite$\}$.*

Theorem 2 *Let $\Phi : X^\omega \to Y^\omega$.*

1. *The domain $dom(\Phi)$ of a DTM-mapping Φ is Π_3-definable. Conversely, for every Π_3-definable set $F \subseteq X^\omega$ there is a DTM M such that $F = dom(\Phi_M)$.*

2. *Let Φ be an SDTM-mapping. Then $dom(\Phi)$ is Π_2-definable. Conversely, for every Π_2-definable set $F \subseteq X^\omega$ there is an SDTM M such that $F = dom(\Phi_M)$.*

Proof.

1. Let Φ be a DTM-mapping, and let φ be a recursive nondecreasing function related to Φ via Theorem 1. Then $\xi \in dom(\Phi)$ if and only if for every $i \in N$ almost all members of the family $(\varphi(w))_{w \in \mathbf{pref}(\xi)}$ have a common prefix of length i, i.e. if and only if $\forall i \exists t \forall n (n \geq t \to \exists v (v \in Y^* \wedge |v| = i \wedge v \sqsubseteq \varphi(\xi/n)))$. Since the quantifier $\exists v$ is a bounded one, the latter formula is a Π_3-condition.

   In order to prove the converse, we use Theorem 9 of [8] which proves that for every Π_3-definable set $F \subseteq X^\omega$ there is an SDTM-mapping $\Psi : X^\omega \to \{a,b\}^\omega$ such that $F = \Psi^{-1}(P_3)$ where $P_3 := \{\xi : \xi \in \{a,b\} \wedge \forall i \exists t \forall n (\xi(\langle i,t,n\rangle) = a)\}$. Here $\langle\cdot,\cdot,\cdot\rangle : N \times N \times N \to N$ is a recursive bijection. For the sake of convenience we assume $\langle\cdot,\cdot,\cdot\rangle$ to be monotone, i.e. $i \leq i', t \leq t'$ and $n \leq n'$ imply $\langle i,t,n\rangle \leq \langle i',t',n'\rangle$.

   In view of Lemma 1 it is sufficient to show that P_3 is the domain of a suitable DTM-mapping Φ_M. For this aim we construct the Turing machine M according to the following program:

   Let $cont(i)$ denote the contents of the i-th cell of the output tape of M. Then this Turing machine M, for every i, changes the contents of the tape cell i, (i.e. $cont(i)$), until it meets a value $t(i)$ such that for all $n \in N$ the input sequence ξ has $\xi(\langle i, t(i), n\rangle) = a$. Otherwise, it will change $cont(i)$ infinitely often. Initially we set $t(0) = 0$ and then, starting with $l = 1$, in the l-th step M does the following:

(a) **read**$(\xi\,(l))$;

(b) $t\,(l) := 0$;

(c) $cont\,(l) := 0$;

(d) **for** all i such that $\langle k, m, n\rangle \le l$ **do**

if $\forall n\,((\langle i, t\,(i)\,, n\rangle \le l) \longrightarrow (\xi\,(\langle i, t\,(i)\,, n\rangle) = a))$

then do not change $cont\,(i)$

else begin change $cont\,(i)$; $t\,(i) := t\,(i) + 1$ **end**

Observe that we can compute k, m, n from $\langle k, m, n\rangle$, hence the loop

for all i such that $\langle k, m, n\rangle \le l$ **do**

includes at most $l + 1$ values, and the condition

$$\forall n\,((\langle i, t\,(i)\,, n\rangle \le l) \longrightarrow (\xi\,(\langle i, t\,(i)\,, n\rangle) = a))$$

has to be checked for at most $l + 1$ values, because the mapping $\langle\cdot,\cdot,\cdot\rangle$ is a bijection.

These explanations show that $\Phi_M(\xi)$ is defined if and only if $\xi \in P_3$.

2. Here the proof may be carried out in a similar way. To be more transparent we derive an alternative proof using Proposition 3 and the methods of [7].

   For a recursive sequential function φ the language $U_\varphi := \{wx : w \in X^* \wedge x \in X \wedge \varphi(w) \neq \varphi(wx)\}$ is recursive, and one easily verifies that $dom(\Phi) = \{\xi : \mathbf{pref}\,(\xi) \cap U_\varphi \text{ is infinite}\}$, where Φ is the SDTM-mapping associated with φ according to Theorem 1.

   Conversely, let $V \subseteq X^*$ be a recursive language and define $\varphi : X^* \to X^*$ as follows:

$$\begin{aligned} \varphi\,(\lambda) &:= \lambda, \\ \varphi\,(wx) &:= \begin{cases} \varphi\,(w)\,, & \text{for } wx \notin V, \\ wx, & \text{for } wx \in V, \end{cases} \end{aligned}$$

   i.e. $\varphi\,(w)$ computes the largest prefix of w in V.

   It is now easy to verify that φ generates an SDTM-mapping Φ which is the identity on $\{\xi : \mathbf{pref}\,(\xi) \cap V \text{ is infinite}\}$ and undefined elsewhere.

q. e. d.

4 Real Functions Defined by Turing Machines

We now regard Turing machines defining consistent mappings $\Phi_M : R_r \to R_s$ as devices for computing real functions.

Remark 1 *It is well-known (see [3], [2]) that the real function Φ_M^R induced by a consistent SDTM-mapping Φ_M is continuous on every closed interval $[a,b] \subseteq dom\left(\Phi_M^R\right)$.*

Remark 2 *The mappings converting expansions at base r to expansions at base s, in general, are not SDTM-mappings (see [1]). On the other hand, it can be shown that for each pair of bases r and s there is a DTM converting every expansion at base r to a corresponding expansion at base s.*

The following proposition shows that every DTM-mapping $\Phi_M : R_r \to R_s$ can be expurgated in such a way that inconsistencies are eliminated.

Proposition 4 *For every DTM M with $\Phi_M : R_r \to R_s$ we can effectively construct a DTM M′ for which $\Phi_{M'} : R_r \to R_s$ is a maximal consistent mapping contained in Φ_M such that*

1. *for all $\xi \in dom\,(\Phi_M)$ with $\mu_r(\xi) \in (R - Q_{r,2})$ we have $\Phi_{M'}(\xi) = \Phi_M(\xi)$, whereas*

2. *for all $\xi \in dom\,(\Phi_M)$ with $\mu_r(\xi) \in Q_{r,2}$ we have $\Phi_{M'}(\xi) = \Phi_M(\xi)$ if and only if $\Phi_M(\xi) = \Phi_M(\xi')$, where $\xi' \neq \xi$ is the second expansion of $\mu_r(\xi)$ (otherwise $\Phi_M(\xi)$ remains undefined).*

Proof. The idea behind the proof is that M' for any input ξ with $\mu_r(\xi) \in Q_{r,2}$ not only computes the digits of the expansion of $\Phi_M(\xi)$, but in parallel also computes the digits of the expansion of $\Phi_M(\xi')$ and compares their values $\mu_s(\Phi_{M'}(\xi))$ and $\mu_s(\Phi_{M'}(\xi'))$.

q. e. d.

The following lemma reveals a severe restriction SDTM-mappings obey to:

Lemma 2 *Let $f : R \to R$ be a continuous real function, which is strictly monotonic from the closed interval $[a,b]$ onto the closed interval $[c,d]$, where $a < b$. If $f^{-1}(y) \notin Q$ for some $y \in Q_{s,2}$, $c < y < d$, then no SDTM M with $\Phi_M : R_r \to R_s$ exists such that Φ_M^R coincides with f on $[a,b]$.*

Proof. Let us assume $0 < c < y < d < 1$ (the changes for the other cases are obvious). Moreover let $y_1 = +0 \cdot y(1)...y(k)0^\omega$ be the terminating expansion of y at base s and let $y_2 = +0 \cdot y(1)...(y(k)-1)(s-1)^\omega$ be the non-terminating expansion of y at base s. Assuming M to be an SDTM computing f on $[a,b]$, M has to write the k-th digit of the fraction of y onto its output tape after having seen $[x]_r \mid m$ of the expansion of x, $x := f^{-1}(y)$, at base r, where $[x]_r = x(0) \cdot x(1)x(2)...$ is the unique expansion of x at base r and $[x]_r \mid m = x(0) \cdot x(1)...x(m)$. As $f^{-1}(y) \notin Q$, two natural numbers m_1 and m_2 with $m < m_1 < m_2$ must exist such that all the following conditions hold:

1. $x(m_1) \neq 0$ and $x(m_2) \neq r-1$.
2. For the two real numbers $x' := x - r^{-m_1}$ and $x'' := x + r^{-m_2}$ we have $a < x' < x < x'' < b$. Obviously, we have $x' \notin Q$ and $x'' \notin Q$.

Assuming, without loss of generality, f to be strictly increasing, we conclude $f(x') < f(x) < f(x'')$. Yet for computing $[f(x')]_s \mid k$ respectively $[f(x'')]_s \mid k$ the Turing machine M could only use $[x']_r \mid m$ respectively $[x'']_r \mid m$, which in both cases coincide with $[x]_r \mid m$. On the other hand

$$f(x') < \mu_s(+0 \cdot y(1)...(y_k - 1)(s-1)^\omega) = \mu_s(+0 \cdot y(1)...y(k)0^\omega) < f(x''),$$

i.e. $[f(x')]_r \mid k$ and $[f(x'')]_r \mid k$ cannot coincide. This contradiction shows that no such SDTM M exists such that Φ_M^R coincides with the given function f on $[a,b]$.

q. e. d.

For example, the preceding lemma shows that well-known mathematical functions like $sin(x)$ or e^x cannot be realized by SDTMs.

Remark 3 *Having explained how one-dimensional real functions can be realized by our Turing machine models, it is straightforward to extend these definitions to more-dimensional real functions as well as to (more-dimensional) complex functions.*

5 Real and Complex Numbers Defined by Turing Machines

Our models of Turing machines also allow us to introduce three different classes of real numbers: Every constant Turing machine mapping $\Phi_M : R_r \to R_s$ defined everywhere (i.e. disregarding the input) specifies the expansion of a unique real number. As the output of the Turing machine M does not depend on the input, in the following we will even assume $dom(\Phi_M) = (B_r \cup \{+,-,\cdot\})^\omega$ instead of restricting $dom(\Phi_M)$ to R_r only. The sets of real numbers defined by SDTMs, DTMs and TMs are denoted by $SDTN(R)$, $DTN(R)$, and $TN(R)$, respectively.

Although, as outlined in Remark 2 and in Lemma 1, on one hand the conversions of expansions at base r to expansions at base s in general cannot be realized by SDTMs and on the other hand DTM-mappings are not closed under compositions, it is easy to see that the sets of real numbers $SDTN(R)$, $DTN(R)$, and $TN(R)$ do not depend on the underlying base r. For any $x \in SDTN(R)$ respectively $x \in DTN(R)$ the expansion of such a real number x can be computed for any arbitrary base s; as $Q \subset SDTN(R) \subset DTN(R)$, we only have to consider two cases:

1. If $x \in Q$, then, for every base s, x even has an expansion at base s that can be computed by a finite automaton [1].
2. If $x \notin Q$, then, for every base s, $x \notin Q_{s,2}$, too. Hence, given x by a(n) (S)DTM $M_{x,r}$ computing an expansion of x at base r, the usual conversion algorithm from base r to base s combined with $M_{x,r}$ yields a(n) (S)DTM $M_{x,s}$ computing an expansion of x at base s.

Similar to Theorem 1, we can obtain characterizations of the classes $SDTN(R)$, $DTN(R)$, and $TN(R)$ defined above. For this aim we have to introduce the first Σ-class of the analytical hierarchy: $F \subseteq X^\omega$ is called Σ_1^1-*definable* if and only if there is a recursive relation $R \subseteq N \times X^* \times X^*$ such that

$$\xi \in F \iff \exists\zeta\forall t\exists n((t, \zeta/n, \xi/n) \in R) .$$

In a similar way, a language $W \subseteq X^*$ is called a Σ_1^1-set if and only if there is a recursive relation $R \subseteq N \times X^* \times X^*$ such that

$$w \in W \iff \exists\zeta\forall t\exists n((t, \zeta/n, w) \in R) .$$

Theorem 3 *Let* $\xi \in R_r$.

1. $\mu_r(\xi) \in SDTN(R) \iff$ **pref** (ξ) *is a recursive set* $\iff \{\xi\}$ *is* Π_1*-definable.*
2. $\mu_r(\xi) \in DTN(R) \iff$ **pref** (ξ) *is a* Π_2*-set as well as a* Σ_2*-set* $\implies \{\xi\}$ *is* Π_2*-definable.*
3. $\mu_r(\xi) \in TN(R) \iff$ **pref** (ξ) *is a* Σ_1^1*-set* $\iff \{\xi\}$ *is* Σ_1^1*-definable.*

Proof.

1. The first equivalence is well-known [3]. In order to show the second equivalence, in view of Proposition 3 we have that $(B_r \cup \{+,-,\cdot\})^* - \mathbf{pref}(\xi)$ is recursively enumerable if and only if $\{\xi\}$ is Π_1-definable. Since $(B_r \cup \{+,-,\cdot\})^* - \mathbf{pref}(\xi)$ contains exactly $(r+3)^l - 1$ words of length l, one easily infers that $B_r^* - \mathbf{pref}(\xi)$ is already recursive.

2. Let Φ_M be the mapping induced by a DTM M and observe that $\{\eta\} = \Phi_M((B_r \cup \{+,-,\cdot\})^\omega)$ if and only if $\eta = \Phi_M(0^\omega)$. From this we obtain $w \in \mathbf{pref}(\eta)$ if and only if there is a t_w such that for all $n \geq t_w$ we have $w \sqsubseteq \varphi_M(0^n)$ where φ_M is a recursive non-decreasing function associated with Φ_M according to Theorem 1. This observation shows that $\mathbf{pref}(\eta)$ is a Σ_2-set. Now Lemma 7.1 of [6] shows that $\mathbf{pref}(\eta)$ is also a Π_2-set which in turn implies that $\{\eta\}$ is Π_2-definable.

   Conversely, let $\mathbf{pref}(\eta)$ be a Σ_2-set, i.e. there is a recursive relation $S \subseteq (B_r \cup \{+,-,\cdot\})^* \times N \times N$ such that $w \in \mathbf{pref}(\eta)$ if and only if $\exists t \forall n\,((w,t,n) \in S)$.

   Define $(w,t,v) \in R :\leftrightarrow (w,t,|v|) \in S$. Then for every $\xi \in (B_r \cup \{+,-,\cdot\})^\omega$ we have $w \in \mathbf{pref}(\eta)$ if and only if $\exists t \forall n((w,t,\xi/n) \in R$. Thus according to Proposition 1 the recursive relation R defines a DTM-mapping $\Phi : (B_r \cup \{+,-,\cdot\})^\omega \rightarrow (B_r \cup \{+,-,\cdot\})^\omega$ with $dom(\Phi) = (B_r \cup \{+,-,\cdot\})^\omega$.

3. Now let us consider the case of a non-deterministic Turing machine M working on the input sequence 0^ω. The non-deterministic choices of M can be described by a sequence $\zeta \in \{0,...,k\}^\omega$, where each number in $\{0,...,k\}$ describes one of all the possible branchings of M. We can construct a DTM M' with input alphabet $\{0,...,k\}$ such that M' outputs $\eta \in (B_r \cup \{+,-,\cdot\})^\omega$ if and only if the sequence $\zeta \in \{0,...,k\}^\omega$ of M' describes a run of M on 0^ω with output η. Then from Theorem 1 it follows that η is the output sequence of a non-deterministic Turing machine M on input 0^ω if and only if $\exists \zeta \forall i\ \exists t\ \forall n\ \ (n \geq t \rightarrow \eta/i \sqsubseteq \varphi(\zeta/n))$ where the prefix of quantifiers according to the Tarski-Kuratowski-algorithm (see [4] or [8]) can be reduced to the form $\exists \zeta \forall j \exists k$. Thus $\{\eta\}$ is Σ_1^1-definable.

   Conversely, if $\{\eta\}$ is Σ_1^1-definable then $\{\eta\}$ is accepted by a non-deterministic Turing machine M'' [8]. From this machine M'' one can construct a machine which guesses an output and checks whether its output equals η. Otherwise it rejects the output by writing infinitely often into a certain cell of the output tape.

q. e. d.

From the characterizations proved in Theorem 3 we immediately obtain

Theorem 4 $\quad Q \subset SDTN(R) \subset DTN(R) \subset TN(R)$.

For concrete examples establishing the strictness of the inclusions in Theorem 4 see [2].

As outlined at the beginning of this section, the sets of real numbers $SDTN(R)$, $DTN(R)$, and $TN(R)$ do not depend on the underlying base. Hence throughout the rest of this paper we shall restrict ourselves to consider expansions of real numbers at base 2 only.

The numbers characterized in Theorem 3 (1) are known as the recursive real numbers of [10]. A recursive real number is usually defined as a real number ρ for which a recursive function $f : N \to Q$ exists such that $|\rho - f(n)| \leq 2^{-n}$. It is now interesting to observe that also the numbers defined by DTMs can be characterized in terms of converging recursive sequences. For the sake of completeness we also reprove this well-known result mentioned above:

Theorem 5 *Let ρ be a real number. Then*

1. *$\rho \in SDTN(R)$ if and only if ρ is recursive;*
2. *$\rho \in DTN(R)$ if and only if there is a recursive function $f : N \to Q$ such that $f(n)$ converges to ρ.*

Proof. If ρ is a rational number, the assertion is obvious. So let ρ be irrational, and without loss of generality we may also assume that $0 < \rho < 1$.

1. Let M be an SDTM which generates an infinite word $\xi \in \{+0\cdot\} B_2^{\omega}$, $\xi = +0 \cdot \xi(1) \xi(2) \ldots$. Then there is a moment at which the generated prefix of ξ on the output tape of M is $+0 \cdot \xi(1) \xi(2) \ldots \xi(n)$. Define $f(0) := +0\cdot$ and, for $n \geq 1$, $f(n) := +0 \cdot \xi(1) \ldots \xi(n)$; obviously, $|\mu_2(f(n)) - \mu_2(\xi)| < 2^{-n}$.

   Conversely, let $f : N \to Q$ define a recursive real number ρ. Then a Turing machine M defining ξ such that $\mu_r(\xi) = \rho$ can be constructed as follows: Successively, for $n = 1, 2, \ldots$ it expands the rational number $f(n)$ up to precision 2^{-n} and considers $v_n := [f(n)]_2 \mid (n-1)$, i.e. the first $n-1$ digits of the fraction of the expansion of the rational number $f(n)$ at base 2. Since $|f(n) - \rho| \leq 2^{-n}$ and $|f(n) - \mu_r(v_n)| \leq 2^{-n}$ we know $|\mu_r(v_n) - \rho| \leq 2^{-(n-1)}$ and thus the first $(n-1)$ letters of the fraction of the (unique) expansion ξ of ρ at base 2 coincide with v_n.

2. Now, let M be a DTM generating an infinite word $\xi \in \{+0\cdot\} B_2^{\omega}$. Let us assume that at the very beginning of its computations M writes $+0\cdot$ onto its output tape and never goes back to these first three cells of its output tape. For $n \geq 1$, let w_n be the prefix $\in \{+0\cdot\} B_2^{*}$ on the output tape up to the first blank symbol after the n-th action of M on its output tape. Define $f(0) := +0\cdot$ and, for $n \geq 1$, $f(n) := w_n$. Since M generates ξ, for every i there is a t_i such that for all $n \geq t_i$ we have $w_n \mid i = \xi \mid i$, hence $|\mu_r(w_n) - \mu_r(\xi)| \leq 2^{-i}$ for $n \geq t_i$. Thus the $\mu_r(w_n)$ converge to $\mu_r(\xi)$.

Conversely, let $f : N \to Q$ be a recursive function such that the $\mu_r(f(n))$ converge to ρ. Define v_n and M as above in the first part of the proof. Then $|\rho - \mu_r(v_n)| \leq |\rho - f(n)| + 2^{-(n-1)}$, and as $f(n)$ converges to ρ, i.e. for all i there is a t_i such that $|\rho - f(n)| \leq 2^{-i}$ for all $n \geq t_i$, the sequence $\mu_r(v_n)$ also converges to ρ. Since ξ is the unique r-ary expansion of ρ it follows that for $n \geq t_i$ the machine M never changes an output letter in ξ/i.

q. e. d.

Similar to Remark 3, the definitions given above can easily be extended to the more-dimensional case as well as to the complex case. The sets of complex numbers defined by SDTMs, DTMs, and TMs are denoted by $SDTN(C)$, $DTN(C)$, and $TN(C)$, respectively.

Remark 4 *Obviously, $z \in SDTN(C)$ ($DTN(C), TN(C)$, respectively) if and only if both $Re(z)$ as well as $Im(z)$ are in $SDTN(R)$ ($DTN(R), TN(R)$, respectively), where $Re(z)$ and $Im(z)$ denote the real and the imaginary part of z ($z = (Re(z), Im(z))$).*

The proof of the following folklore result for strict deterministic Turing machines will also allow us to give an easy proof of the corresponding result for non-deterministic Turing machines.

Lemma 3 *$SDTN(R)$ is a real field, i.e. for any real numbers x and y in $SDTN(R)$, $x + y$, $-y$, $x * y$, and x/y are in $SDTN(R)$, too.*

Proof. If the resulting number $x + y$, $-y$, $x * y$, or x/y is rational, nothing has to be proved any more, because according to Theorem 4 we know that $Q \subset SDTN(R)$. Thus in the following we will assume the resulting number z to be irrational, i.e. $z \notin Q_{2,2}$, and therefore $z = z(0) \cdot z(1) z(2) \ldots$, where for all $m \geq 1$ there exist m', m'' with $m < m' < m''$ such that $z(m') = 0$ and $z(m'') = 1$. Moreover let M_x and M_y be two SDTMs computing the expansions of x respectively y at base 2.

1. An SDTM M_{x+y} computing $[x+y]_2$, i.e. the expansion of $x + y$ at base 2, for $x, y > 0$, can easily be constructed from M_x and M_y in the following way: If $x \in Q_{2,2}$ we can assume M_x to compute the terminating expansion of x, hence the construction is obvious (the same argument holds for $y \in Q_{2,2}$). Therefore let us assume $x \notin Q_{2,2}$ and $y \notin Q_{2,2}$.

   As $x + y \notin Q_{2,2}$, we know that for all $k \geq 1$ there is some $m \geq k$ such that $[x]_2(m) = [y]_2(m)$. The prefix $[x+y]_2 \mid (m-1)$ then is uniquely determined by $f(m) \mid (m-1)$, where $f(m)$ is the (terminating) expansion of

$$\left(\sum_{i=0}^{m-1} \left([x]_2(i) + [y]_2(i)\right) 2^{-i}\right) + [x]_2(m)\, 2^{-(m-1)}.$$

2. For $1 > x > y > 0$, where we again assume $x, y, x-y \notin Q_{2,2}$, an SDTM M_{x-y} computing the unique expansion of $x-y$ at base 2, can be constructed as follows:

   According to our assumptions,

$$x = \sum_{i=1}^{\infty} [x]_2(i)\, 2^{-i} > \sum_{i=1}^{\infty} [y]_2(i)\, 2^{-i} = y$$

   Thus, for any $k \geq 1$ there is some $m \geq k$ such that $[x]_2(m) \neq [y]_2(m)$. $[x-y]_2 \mid (m-1)$ is uniquely determined by $f(m) \mid (m-1)$, where $f(m)$ is the (terminating) expansion of

$$\left(\sum_{i=1}^{m-1} [x]_2(i)\, 2^{-i}\right) - \left(\left(\sum_{i=1}^{m-1} [y]_2(i)\, 2^{-i}\right) + [y]_2(m)\, 2^{-(m-1)}\right).$$

   The remaining cases for the sum $x+y$ for arbitrary $x, y \in SDTN(R)$ are obvious from the constructions given above. Changing the sign of an expansion is trivial, hence so far we conclude that *SDTN(R)* is a group with respect to addition.

3. An SDTM M_{x*y}, for $x * y \notin Q_{2,2}$, $x, y > 0$, successively for $k = 1, 2, \ldots$ computes the (terminating) expansions of

$$z_k^- := \left(\sum_{i=0}^{k} [x]_2(i)\, 2^{-i}\right) * \left(\sum_{j=0}^{k} [y]_2(j)\, 2^{-j}\right)$$

   and

$$z_k^+ := \left(2^{-k} + \sum_{i=0}^{k} [x]_2(i)\, 2^{-i}\right) * \left(2^{-k} + \sum_{j=0}^{k} [y]_2(j)\, 2^{-j}\right).$$

   Obviously, for all $k \geq 1$, $z_k^- < z < z_k^+$ and

$$0 < z_k^+ - z_k^- < 2^{-k}\left(x + y + 2^{-k}\right) < 2^{-k}\left(x + y + 1\right).$$

   This implies that for all $i \geq 1$ there must be an $m \geq 1$ such that

$$\left[z_k^-\right]_2(i) \neq \left[z_k^+\right]_2(i) \text{ for } k < m \text{ and}$$
$$\left[z_k^-\right]_2(i) = \left[z_k^+\right]_2(i) = [z]_2(i) \text{ for } k \geq m,$$

   which recursively determines $[z]_2(i)$.

4. Let $y \in SDTN(R) - Q$, $y > 1$. Then $0 < 1/y < 1$; moreover, $\mu_2([1/y]_2 \mid m) < 1/y < \mu_2([1/y]_2 \mid m) + 2^{-m}$ for all $m \geq 1$, and for each m an $l \leq 1$ must exist such that

$$\mu_2([1/y]_2 \mid m) * (\mu_2([y]_2 \mid l) + 2^{-l}) < 1 \text{ and}$$
$$1 < (\mu_2([1/y]_2 \mid m) + 2^{-m}) * \mu_2([y]_2 \mid l).$$

An SDTM $M_{1/y}$ computing $[1/y]_2$ therefore can proceed as follows in order to compute $[1/y]_2(m)$, $m = 1, 2, \ldots$:

In parallel, for $[1/y]_2(m) = 1$ and for $[1/y]_2(m) = 0$, $M_{1/y}$ computes $\mu_2([1/y]_2 \mid m) * (\mu_2([y]_2 \mid l) + 2^{-l})$ and $(\mu_2([1/y]_2 \mid m) + 2^{-m}) * \mu_2([y]_2 \mid l)$ until the condition given above holds, which uniquely determines the value of $[1/y]_2(m)$.

The remaining cases are obvious and therefore left to the reader.

q. e. d.

Corollary 1 *$TN(R)$ is a real field.*

Proof. Let two real numbers x, y, $x > y > 0$, be given by TMs M_x and M_y. Then, for $\circ \in \{+, -, *, /\}$, TMs $M_{x \circ y}$ computing $x \circ y$ can be constructed according to the formulas derived in the proof of Lemma 3 (provided $z \notin Q$):

The TMs $M_{x \circ y}$ in a non-deterministic way choose the right moments to assume certain prefices of the expansions of x respectively y never to be changed any more and to use these prefices for computing the prefices of $x \circ y$ according to the formulas derived in the proof of Lemma 3. Such runs of $M_{x \circ y}$ obviously yield the unique expansion of $x \circ y$. In all other cases, where $M_{x \circ y}$ during the simulations of M_x respectively M_y would have to change a digit of the expansions of x respectively y that has been assumed to be fixed already, $M_{x \circ y}$ halts (without computing an infinite word on its output tape).

q. e. d.

The closure of $DTN(R)$ with respect to the arithmetic operations can be proved by using the characterization of the real numbers in $DTN(R)$ according to Theorem 5 (2).

Lemma 4 *$DTN(R)$ is a real field.*

Proof. For $x, y \in DTN(R)$, let $f_x : N \to Q$ and $f_y : N \to Q$ be two recursive functions associated with x and y, respectively, according to Theorem 5 (2). Then, for $\circ \in \{+, -, *, /\}$, the function $f_{x \circ y} : N \to Q$ defined by $f_{x \circ y}(n) := f_x(n) \circ f_y(n)$ is recursive and

$$\lim_{n \to \infty} f_{x \circ y}(n) = \lim_{n \to \infty} (f_x(n) \circ f_y(n)) = \left(\lim_{n \to \infty} f_x(n)\right) \circ \left(\lim_{n \to \infty} f_y(n)\right) = x \circ y,$$

hence, according to Theorem 5 (2), $x \circ y \in DTN(R)$, too.

q. e. d.

Theorem 6 *$SDTN(C)$, $DTN(C)$, and $TN(C)$ are subfields of C.*

Proof. The stated results are immediate consequences of Remark 4, Lemma 3, Lemma 4, and Corollary 1.

q. e. d.

The proof of the following result for strict deterministic Turing machines (which for real numbers in the unit interval $[0,1]$ can already be found in [2] or [11]) will also allow us to give an easy proof of the corresponding result for non-deterministic Turing machines as well as for (non-strict) deterministic Turing machines.

Lemma 5 *Let $p(x)$ be a real polynomial of degree n such that all its $n+1$ real coefficients are in $SDTN(R)$. Then each real root of $p(x)$ is in $SDTN(R)$, too.*

Proof. The proof is carried out by induction on n, the degree of the real polynomial p, $p(x) = \sum_{i=0}^{n} a_i x^i$, where $a_i \in SDTN(R)$ for all i with $0 \leq i \leq n$.

1. $n = 1$: $p(x) = a_1 x + a_0$ yields the single root $x_1 = (-a_0)/a_1$. According to Lemma 3, $x_1 \in SDTN(R)$.
2. $n > 1$: Let x_1 be one of the n roots of $p(x)$. If $x_1 \in Q$, then obviously nothing has to be proved any more. Hence in the following we shall assume $x_1 \notin Q$ (and therefore $x_1 \notin Q_{2,2}$, too). We now consider the first derivative $p'(x)$ of the polynomial $p(x)$. Obviously, $p'(x) = \sum_{i=1}^{n} a_i i x^{i-1}$ is a polynomial of degree $n-1$, $n-1 \geq 1$, with coefficients in $SDTN(R)$.
   (a) If x_1 is a root of $p(x)$ such that x_1 is also a root of $p'(x)$, then the induction hypothesis already tells us that $x_1 \in SDTN(R)$.
   (b) If $p'(x_1) \neq 0$, then x_1 is a single root of p, and without loss of generality we shall only consider the case $p'(x_1) > 0$. Then we can choose an $m \geq 1$ and a $q_0 \geq 1$ such that the following conditions hold true:
      i. x_1 is the only root of p in $[q_0 2^{-m}, (q_0+1) 2^{-m}]$, and, moreover, $q_0 2^{-m} \notin N$.
      ii. $p'(x) > 0$ for all $x \in [q_0 2^{-m}, (q_0+1) 2^{-m}]$.
      iii. If $\tilde{p}$ is a polynomial with coefficients in $Q_{2,2}$ such that $\tilde{p}(x) = \sum_{i=0}^{n} \tilde{c}_i x^i$ with $|\tilde{c}_i - c_i| \leq 2^{-m}$ for all i with $0 \leq i \leq n$, then $(\tilde{p})'(x) > 0$ for all $x \in [q_0 2^{-m}, (q_0+1) 2^{-m}]$.

Now starting from this interval $[q_0 2^{-m}, (q_0+1)\,2^{-m}]$ we can construct an infinite sequence of intervals $\left[q_k 2^{-(m+k)}, (q_k+1)\,2^{-(m+k)}\right]$ such that

$$\{x\} = \bigcap_{k=0}^{\infty} \left[q_k 2^{-(m+k)}, (q_k+1)\,2^{-(m+k)}\right]$$

and each step of the bisection algorithm described in the following yields a new digit of the expansion of x_1 at base 2:

Let us define

$$sign\,(y) = \begin{cases} 1 & \text{for } y > 0, \\ 0 & \text{for } y = 0, \\ -1 & \text{for } y < 0 \end{cases}$$

as well as

$$\delta^+(y) = \begin{cases} 1 & \text{for } y > 0, \\ 0 & \text{for } y \leq 0 \end{cases}$$

and

$$\delta^-(y) = \begin{cases} -1 & \text{for } y < 0, \\ 0 & \text{for } y \geq 0 \end{cases}$$

for all $y \in R$.

Moreover, for all i with $0 \leq i \leq n$, let $b_i = +b_i(0)\,.b_i(1)\,b_i(2)\ldots$ be an expansion of $|a_i|$ at base 2. In the following we will not distinguish between $b_i(0)$ and $\mu_2(+b_i(0))$. Given an interval

$$\left[q_k 2^{-(m+k)}, (q_k+1)\,2^{-(m+k)}\right],$$

for testing

$$sign\left(p\left((2q_k+1)\,2^{-(m+k+1)}\right)\right)$$

we successively compute the integer numbers

$$\sum_{i=0}^{n}\left(\left(sign\,(a_i)\sum_{j=0}^{l} b_i(j)\,2^{l-j} + \delta^+(a_i)\right)(2q_k+1)^i\,2^{(n-i)(m+k+1)}\right)$$

and

$$\sum_{i=0}^{n}\left(\left(sign\,(a_i)\sum_{j=0}^{l} b_i(j)\,2^{l-j} + \delta^-(a_i)\right)(2q_k+1)^i\,2^{(n-i)(m+k+1)}\right)$$

for $l = m+k+2, m+k+3, \ldots$ until both numbers are > 0 respectively < 0.

The next interval $\left[q_{k+1} 2^{-(m+k+1)}, (q_{k+1}+1)\,2^{-(m+k+1)}\right]$ with $q_{k+1} = 2q_k$ respectively $q_{k+1} = 2q_k + 1$ is chosen according to the result of these computations in such a way that the evaluations of the polynomial in the bounds of the interval have different signs, i.e. $sign\left(p\left(q_{k+1}2^{-(m+k+1)}\right)\right) \neq sign\left(p\left((q_{k+1}+1)\,2^{-(m+k+1)}\right)\right)$. Obviously, the steps of this algorithm described above can be carried out by an SDTM M, and each such step of M yields a new digit of the expansion of x_1 from (the terminating expansion of) $q_k 2^{-(m+k)}$.

q. e. d.

Similar arguments like those used in the proof of Corollary 1 show that the same constructions as elaborated in the proof of Lemma 5 can be taken over for proving the algebraic closure of $TN(R)$, too.

Corollary 2 *Let $p(x)$ be a real polynomial of degree n such that all its $n+1$ real coefficients are in $TN(R)$. Then each real root of $p(x)$ is in $TN(R)$, too.*

Proof. In the case of polynomials with coefficients in $TN(C)$, the moments for assuming certain prefices of the coefficients never to be changed any more can be chosen in a non-deterministic way again as outlined in the proof of Corollary 1.

q. e. d.

Corollary 3 *Let $p(x)$ be a real polynomial of degree n such that all its $n+1$ real coefficients are in $DTN(R)$. Then each real root of $p(x)$ is in $DTN(R)$, too.*

Proof. In the case of polynomials with coefficients in $DTN(R)$, we use a working tape of M for storing the current value of the prefix of x_1 computed so far by using the current prefices of the coefficients computed by simulating the corresponding DTMs. When one of the DTMs for the coefficients rewrites a digit previously used for the computations of the current prefix of the root x_1, then M has to redo the computations by using the new prefices of the coefficients. Changes to the output tape of M are only taken over from the working tape mentioned above, when additional digits are written there.

q. e. d.

Theorem 7 *$SDTN(C)$, $DTN(C)$, and $TN(C)$ are algebraically closed subfields of C.*

Proof. As is well-known from algebra, for any real field K, $K \subseteq R$, the algebraic closure of K in C, $\overline{K}$, has the following properties: For all $\alpha, \beta \in R$, we have $(\alpha, \beta) \in \overline{K}$ if and only if for α as well as for β a real polynomial p_α

respectively p_β with coefficients in K exists such that α is a real root of p_α and β is a real root of p_β.

As we have shown in Lemma 5, Corollary 3, and Corollary 2, for each $K(R)$ with $K \in \{SDTN, DTN, TN\}$ every real root of any real polynomial with coefficients in K again is in $K(R)$; hence, according to the observations above and our definitions of the sets $K(C)$ (see Remark 4) we have

$$\overline{K(R)} = \left(\overline{K(R)} \cap R\right) \times \left(\overline{K(R)} \cap R\right) = K(R) \times K(R) = K(C).$$

Thus we obtain $K(C) = \overline{K(R)}$ for each K in $\{SDTN, DTN, TN\}$, which completes the proof.

q. e. d.

References

[1] S. Eilenberg: *Automata, Languages and Machines, Vol. A* (Academic Press, New York, 1974).

[2] R. Freund: *Real Functions and Numbers Defined by Turing Machines*, Theoret. Comp. Sci. **23** (1983) 287-304.

[3] M. L. Minsky: *Berechnung: Endliche und unendliche Maschinen* (Verlag Berliner Union, Stuttgart, 1971).

[4] H. Rogers: *Theory of Recursive Functions and Effective Computability* (McGraw Hill, 1967).

[5] A. Salomaa: *Formal Languages* (Academic Press, New York, 1973).

[6] L. Staiger: *Hierarchies of Recursive ω-Languages*, J. Inform. Process. Cybernet. EIK **22** (1986) 5/6, 219-241.

[7] L. Staiger: *Sequential Mappings of ω-Languages*, RAIRO Infor. Théor. Appl. **21** (1987) 2, 147-173.

[8] L. Staiger and K. Wagner, *Rekursive Folgenmengen* I, Z. Math Logik Grundlag. Math. **24** (1978) 523-538.

[9] K. Wagner: *Arithmetische Operatoren*, Z. Math Logik Grundlag. Math. **22** (1976) 553-570.

[10] K. Weihrauch: *Computability* (Springer-Verlag, Berlin, 1987).

[11] K. Weihrauch: *A simple Introduction to Computable Analysis*, Technical Report 171, Fernuniversität Hagen, 1995.

[illegible] $\mathfrak{H}$ [illegible] exists such that [illegible] root of [illegible] and [illegible] rank of [illegible]

As we have shown in Lemma [illegible] and Corollary [illegible] [illegible] [illegible] every [illegible] root of any [illegible] polynomial with coefficients in K [illegible] [illegible] [illegible] observations [illegible] and [illegible] [illegible] we [illegible]

$$[illegible]$$

Thus [illegible] for each [illegible] which completes the proof. [illegible]

References

[1] [illegible] Algorithms [illegible] (New York, [illegible])

[2] [illegible] [illegible] 23 (19[illegible]) [illegible]

[3] [illegible] [illegible] [illegible]

[4] [illegible]

[5] [illegible]

[6] [illegible] 32 (19[illegible]) [illegible] 241

[7] [illegible] RAIRO [illegible]

[8] [illegible] 83 (19[illegible]) [illegible]

[9] [illegible] Computing [illegible] [illegible] (19[illegible]) [illegible]

[10] [illegible] Springer-Verlag, Berlin 1991.

[11] [illegible] A Simple Introduction to Computer [illegible] Lecture Notes in Computer Science [illegible] 198[illegible]

The Executive Board of the Kurt Gödel Society

Call for Papers and Instructions to Authors

Besides invited papers and articles directly communicated by the Editorial Board, the Annals of the KGS also publish submitted research or survey papers, as well as short research notes and reviews.

The Annals welcome contributions in the areas of mathematical logic, philosophical logic, computational logic, theoretical computer science, philosophy of physics, knowledge representation, history of logic, and related fields.

Papers submitted for publication must be original and should not be submitted elsewhere. Three copies of the submitted paper should be sent to the Editor in Chief or to any appropriate member of the Editorial Board. The Editorial Board consists of the members of the Executive Board of the Kurt Gödel Society, the Editor in Chief is Alexander Leitsch, the Managing Editor is Katrin Seyr.

Preferably, manuscripts should be written in English, but French and German are admitted too. Manuscripts should be formatted in LaTeX with the macro package llncs.sty available from http://logic.tuwien.ac.at/kgs/annals.html or by e-mail from kgs@logic.tuwien.ac.at. Otherwise, final versions of accepted manuscripts should comply with the format instructions available from the Kurt Gödel Society, Technische Universität Wien, Institut für Computersprachen E 185/2, Resselgasse 3/I/3, A-1040 Wien.

Electronic submissions in postscript format are also welcome. They should be sent to leitsch@logic.tuwien.ac.at with an accompanying email explaining to which member of the Editorial Board they are directed.

All submissions are refereed by peer scientists. It is understood that the submission of a paper is approved by all its authors and that all authors of the paper agree to grant the copyright for this paper in its widest sense to the publishing company publishing the Annals of the KGS.

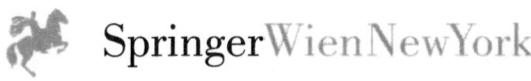